Abstracts and Indexes in Science and Technology
A Descriptive Guide

by
Dolores B. Owen

Second Edition

The Scarecrow Press, Inc.
Metuchen, N.J., and London
1985

Library of Congress Cataloging in Publication Data
Owen, Dolores B.
  Abstracts and indexes in science and technology.
   1. Science--Abstracts--Periodicals--Indexes.
 2. Science--Bibliography--Periodicals--Indexes.
 3. Technology--Abstracts--Periodicals--Indexes.
 4. Technology--Bibliography--Periodicals--Indexes.
I. Title
Z7403.O95   1985   [Q158.5]   016.5   84-10902
ISBN 0-8108-1712-8

Copyright © 1985 by Dolores B. Owen
Manufactured in the United States of America

To my darling daughters,

Sandy, Gabi, and Monique

PREFACE

The purpose of this book, as the title reflects, is to provide a description of the materials one would encounter in conducting a literature search in science or technology. It is by no means intended to be a definitive bibliography, nor is an attempt made to analyze all the idiosyncrasies of every index or abstracting tool included. Instead, it is hoped to facilitate the use of these materials by means of an outline which provides an introduction to their use, gives some indication of their content, and presents researchers with an idea of what they might expect from them. Some of the works, particularly several of those carried over from the first edition, have ceased publication, but most of the material is current. Coverage has been increased in this revision from 125 to 223 titles.

All components of the outline do not always fit the publications included. When this occurs, that segment is omitted, except in the case of "Abstracts." In these instances (almost all indexes, for example), it is stated that this term is not applicable and a description is given of the type of entries found in the publications.

A broad subject arrangement has been utilized, and within the subjects the titles are arranged alphabetically. The choice of a subject for a particular title may not always concur with that which the user would adopt. There are several examples of this, but it is particularly conspicuous of anthropological material placed with "Earth Science." This assignment was based on the premise that physical anthropology fits more comfortably in this category than in any other. If the researcher fails to find an entry where expected, the "Index" or "Table of Contents" will guide him to its location.

The Index includes current titles, variant or superseded titles, and in most cases entries for parts or sections of a publication if those parts have a specific or familiar title.

There are no index entries for country except for U.S. <u>Government Reports Announcements</u> and its variations. Institutions, organizations, bureaus, or departments are listed, when they help guide the user to a certain title. There are also index entries for miscellaneous material found in the text.

A new element has been added to the outline in this edition to accommodate information found in databases. With the advent of online bibliographic searching and its ever-widening availability, this aspect of a bibliographic source seems essential. The name of any available databases has been supplied, the beginning date of a machine readable version of an index or abstract journal is given, and the frequency of updates and the availability of keywords and abstracts is provided whenever possible. Information concerning vendors and prices for the various databases has not been given, principally because of continual change.

The first edition of this title was produced because my then co-author, Marguerite Hanchey and I felt the need for such a description. Marguerite died in 1974 before final publication of that work. This second edition comes about by a similar need for a revision and an updating of the original book. My only regret has been that I have not had her encouragement and guidance to help see it through.

I wish to thank my friends at the Troy H. Middleton Library at Louisiana State University in Baton Rouge for their help and encouragement in this endeavor, my husband, Travis, for his patience, Dru Menard for her typing, Cindy Rice for her ruined eyesight acquired by proofreading, and the Interlibrary Loan Librarians of the world for their efforts in my behalf. I am grateful also to the publishers and editors of the material included for their help and advice with the descriptions. Material from H. W. Wilson publications was reproduced by permission of that company.

INTRODUCTION

Following is an example of the outline devised to explain the titles included in this description, with an indication of what each segment of that outline is intended to convey.

Arrangement: Describes how the material is presented in the various publications (by subject, title, author, or by some other method).

Coverage: Gives an indication of the subject matter contained in the publications, and sometimes extends to the type of material (patents, symposia, books, etc.).

Scope: Shows the origin of the publications covered; the term "International" designates worldwide coverage.

Locating Material: Tells how to access the information contained in the publications.

Abstracts: Discusses the entries one will encounter and gives details of the abstracts if they occur.

Indexes: Gives the user an idea of what to expect from the indexes in the publications and relates their frequency, type, cumulations, etc.

Other Material: Explains features in the publications not covered by other segments of the outline.

viii / Introduction

Periodicals Scanned: Indicates that a list of the periodicals used in compiling the abstracts and indexes is provided, and gives its location and frequency if possible.

Databases: Notes the availability of online sources for the printed material.

Each main entry in this description denotes the current title of the publication, its beginning date (and ending date if it has ceased), the publisher, frequency and a brief bibliographic history if this seems important. This history includes such items as superseded titles, forerunners and predecessors.

It is hoped that this format will prove most useful for the researcher seeking information, and facilitate the librarian's explanation of how it can be obtained.

TABLE OF CONTENTS

Preface v

Introduction vii

GENERAL

Air University Library Index to Military Periodicals	1
Antarctic Bibliography	2
Applied Science and Technology Index	3
Bulletin Signalétique	4
Current Contents	6
Dissertation Abstracts International, Section B, The Sciences and Engineering	7
General Science Index	8
Government Reports Announcements	9
Index to Scientific and Technical Proceedings	10
Information Science Abstracts	12
International Abstracts in Operation Research	13
Masters Theses in the Pure and Applied Sciences Accepted by Colleges and Universities in the United States and Canada	14
Pandex Current Index to Scientific and Technical Literature	14
Proceedings in Print	15
Referativnyi Zhurnal	16
Science Abstracts	19
Science Citation Index	20
Technical Book Review Index	21
Translations Register Index	22
World Transindex	23

MATHEMATICS, STATISTICS, AND COMPUTER SCIENCE

Computer Abstracts	25

Computer and Information Systems
   Abstracts Journal 26
Computer Program Abstracts 27
Computing Reviews 28
Mathematical Reviews 29
New Literature on Automation 30
Statistical Theory and Method Abstracts 30
Zentralblatt für Mathematik und ihre
   Grenzgebiete 31

ASTRONOMY

Astronomischer Jahresbericht 33
Astronomy and Astrophysics Abstracts 34
Meteorological and Geoastrophysical Abstracts 36

CHEMISTRY AND PHYSICS

Acoustics Abstracts 39
Analytical Abstracts 39
Biochemistry Abstracts 40
Chemical Abstracts 41
Chemisches Zentralblatt 43
Chemo-Reception Abstracts 44
Current Abstracts of Chemistry and Index
   Chemicus 45
Current Chemical Reactions 46
Current Physics Index 46
Electroanalytical Abstracts 47
Journal of Current Laser Abstracts 48
Physikalische Berichte 49
Rheology Abstracts, a Survey of World
   Literature 49
Science Research Abstracts Journal 50
Solid State Abstracts Journal 51
Spectrochemical Abstracts 52
Theoretical Chemical Engineering Abstracts 52

NUCLEAR SCIENCE AND SPACE SCIENCE

INIS Atomindex: an International Abstracting
   Service 55
International Aerospace Abstracts 56
Nuclear Science Abstracts 57

Scientific and Technical Aerospace Reports	58

EARTH SCIENCES, ARCHAEOLOGY, AND ANTHROPOLOGY

Abstracts in Anthropology	61
Abstracts of North American Geology	61
Bibliographie Géographique Internationale	62
Bibliography and Index of Geology	63
Bibliography and Index of Micropaleontology	64
Bibliography of North American Geology	65
Bibliography of Seismology	66
British Archaeological Abstracts	67
British Geological Literature	67
Cadmium Abstracts	68
Geo Abstracts	69
Geophysical Abstracts	70
Geophysics and Tectonics Abstracts	71
IMM Abstracts	72
International Petroleum Abstracts	72
Lead Abstracts	73
Metals Abstracts	74
Mineralogical Abstracts	75
Petroleum Abstracts	75
Zentralblatt für Geologie und Paläontologie	76
Zentralblatt für Mineralogie	77
Zinc Abstracts	78

ENGINEERING AND TECHNOLOGY

Applied Mechanics Reviews	79
Bioengineering Abstracts	79
Communication Abstracts	80
Corrosion Abstracts	81
Current Technology Index	81
Design Abstracts International	82
Electronics and Communication Abstracts Journal	83
Engineering Index	84
Gas Abstracts	85
Highway Safety Literature	86
HRIS Abstracts	87
Offshore Abstracts	88
Road Abstracts	88

Safety Science Abstracts Journal   89
Transportation Research Abstracts   90
Urban Mass Transportation Abstracts   91

ENERGY AND ENVIRONMENT

Air Pollution Abstracts, 1932-1972   93
Air Pollution Abstracts, 1970-1976   93
Ecology Abstracts   94
Energy Abstracts for Policy Analysis   95
Energy Information Abstracts   96
Energy Research Abstracts   97
Environment Abstracts   98
Environmental Quality Abstracts   100
Fuel and Energy Abstracts   100
Hydata   101
Land Use Planning Abstracts   102
Pesticides Abstracts   103
Pesticides Documentation Bulletin   104
Pollution Abstracts   104
Selected Water Resources Abstracts   105
Water Pollution Abstracts   107

BIOLOGICAL SCIENCES

Animal Behavior Abstracts   109
Apicultural Abstracts   110
Aquatic Sciences and Fisheries Abstracts   111
Bibliography of Bioethics   112
Bibliography of Reproduction   113
Biological Abstracts   114
Carbohydrate Chemistry and Metabolism
    Abstracts   117
Current Advances in Genetics   117
Current Advances in Plant Science   118
Deepsea Research   119
Entomology Abstracts   120
Excerpta Botanica   121
Genetics Abstracts   122
Helminthological Abstracts   123
Immunology Abstracts   124
Index to American Botanical Literature   124
International Abstracts of Biological
    Sciences   125
Keyword Index of Wildlife Research   126

Marine Fisheries Abstracts	127
Microbiology Abstracts	128
Oceanic Abstracts	129
Protozoological Abstracts	130
Review of Applied Entomology	131
Review of Plant Pathology	132
Sport Fishery Abstracts	133
Tissue Culture Abstracts	133
Wildlife Review	134
World Fisheries Abstracts	135
Zoological Record	136

AGRICULTURAL SCIENCES

Abstracts on Tropical Agriculture	139
Agricultural Engineering Abstracts	139
Agricultural Literature of Czechoslovakia	140
Agrindex	141
American Bibliography of Agricultural Economics	142
Animal Breeding Abstracts	143
Arid Lands Development Abstracts	144
Bibliography of Agriculture	144
Biological and Agricultural Index	146
Cotton and Tropical Fibres Abstracts	146
Crop Physiology Abstracts	147
Dairy Science Abstracts	148
Faba Bean Abstracts	149
Farm and Garden Index	150
Fertilizer Abstracts	150
Field Crops Abstracts	151
Food Science and Technology Abstracts	152
Forest Products Abstracts	153
Forestry Abstracts	153
Herbage Abstracts	154
Home Economics Research Abstracts	155
Horticultural Abstracts	156
Irrigation and Drainage Abstracts	157
Maize Quality and Protein Abstracts	158
Nutrition Abstracts and Reviews	159
Nutrition Planning	160
Ornamental Horticulture	160
Pig News and Information	161
Plant Breeding Abstracts	162
Plant Growth Regulator Abstracts	163
Potato Abstracts	164

Poultry Abstracts	165
Rice Abstracts	166
Seed Abstracts	166
Soils and Fertilizer	167
Sorghum and Millets Abstracts	168
Soyabean Abstracts	169
Textile Technology Digest	170
Tobacco Abstracts	170
Triticale Abstracts	171
Tropical Oil Seeds Abstracts	172
Weed Abstracts	173
World Agricultural Economics and Rural Sociological Abstracts	174
World Textile Abstracts	175

HEALTH SCIENCES

Abstracts of Health Care Management Studies	177
Abstracts of World Medicine	178
Abstracts on Hygiene and Communicable Diseases	178
Accumulative Veterinary Index	179
Animal Disease Occurrence	179
British Medicine	180
Calcified Tissue Abstracts	181
Cancer Therapy Abstracts	182
Cumulative Index to Nursing and Allied Health Literature	183
Current Literature in Family Planning	184
Current Literature on Aging	184
Dental Abstracts	185
Developmental Disabilities Abstracts	186
Drug Abuse Bibliography	186
DSH Abstracts	187
Excerpta Medica	188
Gerontological Abstracts	190
Hospital Abstracts	190
Hospital Literature Index	191
Index Catalogue of Medical and Veterinary Zoology	192
Index Medicus	194
Index Veterinarius	196
International Nursing Index	197
International Pharmaceutical Abstracts	198
Leukemia Abstracts	199

Meditsinskii Referativnyi Zhurnal	200
Mental Health Book Review Index	200
Multiple Sclerosis Indicative Abstracts	201
Neurosciences Abstracts	202
Oral Research Abstracts	202
Physical Education Index	203
Physical Fitness/Sports Medicine	204
Psychological Abstracts	204
Psychopharmacology Abstracts	206
Public Health Engineering Abstracts	206
Rehabilitation Literature	207
Review of Medical and Veterinary Mycology	207
Sleep Bulletin	208
Small Animal Abstracts	209
Toxicology Abstracts	210
Tropical Diseases Bulletin	210
Veterinary Bulletin	211
Virology Abstracts	212
Wildlife Disease Review	213
Index	215

GENERAL

AIR UNIVERSITY LIBRARY INDEX TO MILITARY PERIODICALS, 1949-
U.S. Air University Library. Quarterly with annual cumulations. (Triennial cumulations until 1952-67). Formerly: Air University Periodical Index.

Arrangement: Alphabetical by author and subject, interfiled. As this is an index of military periodicals one is bound to see an abundance of military terms.

Coverage: Subject matter is varied and includes material other than scientific and technical information, e.g., economics, military law, and history. One may expect to find such terms as aeronautical research, communication systems, hydrography, navigation, and space as subject entries.

Scope: Limited to English language military periodicals. It is specifically stated that those included are not generally found in commercial indexing services.

Locating Material: The subject terms are arranged alphabetically with agencies listed directly under their names as subjects. Authors' names are interfiled with subjects. Military services of other countries are listed by the type of service, subdivided by country. The searcher will find book reviews grouped together under that term as a subject. The entries are listed alphabetically by title under each subject entry. There are "see" and "see also" references to alternative terms. All articles by an author are found listed alphabetically under his name and under the proper subject entry. Complete citations are given in both places.

Abstracts: Not applicable. Entries list title, author and journal citation. Illustrative material is noted.

Other Material: Book reviews are found grouped together under that term. A brief statement covering the arrangement and an explanation of the entries are found in the preliminary pages of each issue. A list of abbreviations and a "key to abbreviations" are provided.

Periodicals Scanned: A list is in each quarterly and annual issue.

ANTARCTIC BIBLIOGRAPHY, 1965-
U.S. Library of Congress, Science and Technology Division (Distributed by the U.S. Government Printing Office). Annual. Material first appears in monthly issues under the title Current Antarctic Literature.

Arrangement: The material is arranged by subject. Items that apply to more than one category will be found in only one section and cross referenced at the end of other pertinent sections.

Coverage: Biology, medical sciences, geology, atmospheric and terrestrial physics, meteorology, oceanography, logistics, political geography, expeditions, cartography, and ice and snow are among the subjects included. The coverage begins in volume one with material from 1962.

Scope: International.

Locating Material: Entries are arranged chronologically by accession number within each subject section. The numbering began with volume one in 1965 and continues through all the volumes. Each subject section is given a letter and this precedes each entry (e.g., A, general; B, biological sciences; C, cartography; D, expeditions; etc.). Index entries refer to these letters and numbers.

Abstracts: Titles are given first in English, then in the language of the article. All authors are listed. The Introduction states that "...as a rule, the abstracts are informative rather than descriptive" and generally author abstracts are used unchanged or altered only for the sake of brevity or conformity to the guidelines of the bibliography. If modification has been made, this is noted. The number of references, the language of the article, and an indication as to whether there is an English summary are given.

Indexes: An author index that includes co-author and journal titles for anonymous articles, a subject index with headings, subheadings, and numerous "see" and "see also" references are provided. A geographic index that gives names of places, stations, and geographic features (approved by the U.S. Board on Geographic Names), and a grantee index which names institutions or organizations that received financial support from the National Science Foundation for abstracts or works that are included are also furnished. The entries refer to the letters for subject sections and accession numbers within each section. A cumulative index to volumes 1-7 appeared in 1977. This contains the same four indexes and references are to volume, subject letter, and accession number. In the author, subject, and geographic indexes, the titles are given in English, but the language of the article is indicated. In the grantee index, reference is made only to volume and entry number.

Other Material: A companion volume, Antarctic Bibliography, 1951-1961 extends the coverage retrospectively for ten years. The earlier period is covered by a work published in 1951 by the U.S. Naval Photographic Interpretation Center titled Antarctic Bibliography which contained 5,500 items arranged by subject with an author index.

Periodicals: No list is provided. Although most of the publications are in the Library of Congress collection, some were made available from other institutions and some were review copies or reprints sent from publishers.

Database: Citations are available through COLD REGIONS from 1962 and are updated quarterly. Abstracts and keywords are provided. The Bibliography on Cold Regions Science and Technology is also included.

APPLIED SCIENCE AND TECHNOLOGY INDEX, (Vol. 46), 1958- H.W. Wilson Company. Monthly except July with quarterly and annual cumulations. Supersedes in part Industrial Arts Index, 1913-1957, and continues its numbering. (Material from Applied Science and Technology Index is reproduced by permission of the H.W. Wilson Company.)

Arrangement: Alphabetical by subject. Occasionally subheadings are found under broad areas. Under each subject heading and subheading the entries are alphabetical by title.

Coverage: Aeronautics, automation, chemistry, electricity and electronics, engineering, geology and metallurgy, industrial and mechanical arts, physics, transportation, etc. A list is provided on the cover of the monthly issues and the quarterly cumulations and in the "Prefatory Note" of the bound volume.

Scope: English language journals. Inclusion is determined by subscriber vote.

Locating Material: The searcher determines the proper subject area and scans the material.

Abstracts: Not applicable. Entries provide complete title, author and joint author, but only the first if there are more than two. Complete bibliographic information is given and illustrative material is noted. Frequent "see" and "see also" references guide the user to alternative terms.

Other Material: Abbreviations used in the entries and for the periodical titles are provided in the preliminary pages of each issue and in the cumulations. There is an alphabetical author (or title if no author is given) list of citations to book reviews at the end of each issue and the bound volumes.

Periodicals Scanned: A list is provided in each issue and in the annual cumulations.

BULLETIN SIGNALETIQUE, 1940-
Centre de Documentation du Centre National de la Recherche Scientifique. Beginning with Vol. 3, each number issued in two parts. From Vol. 22, 1961, issued in various sections and combined sections, each having an assigned number. The titles and frequency of the sections vary. Presently published monthly in more than 60 sections. Formerly: Bulletin Analytique, 1940-1955.

Following is a list of sections and their numbers:

101 Sciences de l'Information--Documentation
110 Informatique--Automatique--Recherche Opérationnelle--Gestion--Economie
120 Astronomie--Physique Spatiale--Géophysique
130 Physique Mathématique, Optique, Acoustique, Méchanique, Chaleur
140 Eldoc--Electrotechnique
145 Eldoc--Electronique
150 Physique, Chimie et Technologie Nucléaires
160 Physique de l'Etat Condensé
161 Structure de l'Etat Condensé. Cristallographie
165 Atomes et Molécules. Plasmas
166 GAP HYOR, Atomes, Molécules Gaz Neutres et Ionsés
170 Chimie
220 Minéralogie--Géochimie--Géologie Extraterrestre (Bulletin Signalétique. Bibliographie des Sciences de la Terre)
222 Roches Cristallines (Bulletin Signalétique. Bibliographie des Sciences de la Terre)
223 Roches Sédmentaires--Géologie Marine (Bulletin Signalétique. Bibliographie des Sciences de la Terre)
224 Stratigraphie--Géologie Régionale et Géologie Génerale (Bulletin Signalétique. Bibliographie des Sciences de la Terre)
225 Tectonique (Bulletin Signalétique, Bibliographie des Sciences de la Terre)
226 Hydrologie--Géologie de l'Ingenieur--Formations Superficielles (Bulletin Signalétique. Bibliographie des Sciences de la Terre)
227 Paléontologie (Bulletin Signalétique, Bibliographie des Sciences de la Terre)
310 Génie Biomédical--Informatique Biomédical
320 Biochimie--Biophysique Moleculaire

General / 5

330 Sciences Pharmacologiques--Toxicologie
340 Microbiologie--Virologie--Immunologie
346 Ophtalmologie
347 Oto-Rhino-Laryngologie--Stomatologie--Pathologie Cervicofaciale
348 Dermatologie--Vénéréologie
349 Anesthésie--Réanimation
351 Revue Bibliographique Cancer (editée par l'Institut Gustave Roussy)
352 Maladies de l'Appareil Respiratoire du Coeur et des Vaisseaux--Chirurgie Thoracique et Vasculaire
354 Maladies de l'Appareil Digestif--Chirurgie Abdominale
355 Maladies des Reins et des Voies Urinaires--Chirurgie de l'Appareil Urinaire
356 Maladies du Système Nerveux--Myopathies--Neurochirurgie
357 Maladies des Os et des Articulations--Chirurgie Orthopédique--Traumatologie
359 Maladies du Sang
361 Reproduction. Embryologie. Endocrinologie
362 Diabète. Obesité. Maladies Métaboliques
363 Génétique
364 Protozoaires et Invertébrés. Zoologie Générale et Appliquée
365 Zoologie des Vertébrés. Ecologie Animale. Physiologie Appliqué Humaine.
370 Biologie et Physiologie Végétales. Sylviculture
380 Produits Alimentaires
381 Sciences Agronomiques. Produits Végétales
390 Psychologie--Psychopathologie--Psychiatrie
519 Philosophie
520 Sciences de l'Education
521 Sociologie--Ethnologie
522 Histoire des Sciences et des Techniques
523 Histoire et Science de la Littérature
524 Sciences du Langage
525 Préhistoire et Protohistoire
526 Art et Archéologie
527 Histoire et Sciences des Religions
528 Bibliographie Internationale de Science Administrative
730 Combustibles--Energie
740 Métaux--Métallurgie
745 Soudage, Brasage et Techniques Connexes
761 Microscopie Electronique--Diffraction Electronique
780 Polymères. Peintures. Bois. Cuirs.
880 Génie Chimique--Industries Chimique et Parachimique
885 Nuisances
891 Industries Mécaniques
892 Bâtiment. Travaux Publics. Transports

Arrangement: By broad subject areas subdivided into smaller topics.

Coverage: Each section covers in depth the subject matter delineated in its title and listed in its Table of Contents.

6 / Abstracts and Indexes

Scope: International.

Locating Material: Abstracts are numbered consecutively through the volume. The number also includes the volume number of the section, the section number and the consecutive number of the abstract. Only the latter is used in the index. For example: 34-101-20 refers to the twentieth abstract in Volume 34 of section 101.

Abstracts: Titles are given in Western European languages and in French. The abstracts are in French. Senior author's location is given and the number of references is indicated. The abstracts are usually brief.

Indexes: An author index is found in each issue. Most sections also contain a subject index. There are annual cumulative author and subject indexes.

Other Material: A list of abbreviations and symbols is in each issue.

Periodicals Scanned: From 1972 an annual list of periodicals received at Centre de Documentation is published as a supplement to Bulletin Signalétique.

CURRENT CONTENTS, 1957-
Institute for Scientific Information. Weekly. The beginning date, 1957, represents the earliest time this service appeared. The title has varied somewhat and the publication ultimately split into a number of sections which change as usage demands to reflect the material covered. The beginning dates for different sections vary. Sections presently being published are: Life Sciences; Physical, Chemical and Earth Sciences; Agricultural, Biological and Environmental Sciences; Clinical Practice; Engineering, Technology, and Applied Sciences; Social and Behavioral Sciences; Arts and Humanities.

Arrangement: Arranged by subject. A list of subject headings is provided in the preliminary pages of each issue.

Coverage: A wide range of heavily used material in each subject field is included.

Scope: International

Locating Material: No location numbers are used. The numbers found in the oval by each entry are for ordering purposes.

Abstracts: Not applicable. This service reproduces the tables of contents of journals. The language of the journal is specified and if articles or abstracts appear in English or in a language other than that of the country of origin, this is noted.

General / 7

Indexes: An author index is in each issue which gives name and address of the first author only. A "weekly subject index" in each issue lists the significant words from every title announced in the issue. Under each word, the page in Current Contents where it may be found is given on the left and the journal page number is found on the right. A triennial journal index is provided which gives the issue numbers and pages where a specific journal's table of contents may be found.

Other Material: A topical essay, "Current Comments" is included in each issue.

Periodicals Scanned: Each issue carries a list of journals for which tables of contents are included with a reference to the page number where they may be found. Also in each issue is a list of journals added or deleted. A trimestrial complete list of journals is given and triennially there is a cumulative list which serves as an index (see above), and a publishers' address list.

Database: Items from Current Contents--Clinical Practice from 1974, from Engineering, Technology, and Applied Science, and from Agriculture, Biology and Environmental Sciences, not presently covered in the printed Science Citation Index, are included in SCISEARCH from 1976.

DISSERTATION ABSTRACTS INTERNATIONAL, Section B: THE SCIENCES AND ENGINEERING, 1938-
Xerox, University Microfilms. Monthly. Previous titles: Microfilm Abstracts; Dissertation Abstracts. With Volume 27, split into sections A and B. (Section A covers Humanities and Social Sciences.)

Arrangement: Arranged by subject. The Table of Contents forms a list with some fields (e. g., Agriculture, Chemistry, and Engineering) subdivided into more narrow disciplines.

Coverage: Doctoral dissertations in the sciences and engineering sent to University Microfilms by cooperating institutions in the United States, Canada, and Europe are included. Not all institutions send all their doctoral dissertations. A list of institutions is provided. Some institutions, at the time of entry, had their earlier dissertations included. A list of subjects covered is found in the preliminary pages.

Scope: Contributing institutions in the United States, Canada, and Europe; only dissertations released to University Microfilms may be abstracted.

Locating Material: No numbering is used. The searcher is referred to the proper subject section and page number. This is no problem as there are usually no more than three titles on any given page.

Abstracts: Titles and abstracts are in English and are usually quite long and detailed. The title, author, the institution, the year the dissertation was completed, and in most cases the supervisor are given. At the end, an order number for obtaining a copy of the dissertation and the number of pages are given.

Indexes: A keyword title index (which lists the references alphabetically by keyword) is provided monthly. Titles in which the keywords occur follow in alphabetical order; in the subject section, author, page number, and section (A or B) are given. An author index is also provided monthly and both cumulate annually. A Cumulative Subject and Author Index to volumes 1-5 was published separately.

Other Material: A description of the substitutions made for titles with subscripts, superscripts, and symbols is found in the preliminary pages. A list of cooperating institutions (beginning with Volume 26) gives the date of entry of each institution. Order information for obtaining copies of dissertations is provided in each issue.

Database: COMPREHENSIVE DISSERTATION INDEX (CDI) provides citations from 1861. There are monthly updates. Abstracts and keywords are provided. American Doctoral Dissertations and Comprehensive Dissertations Index are also included.

GENERAL SCIENCE INDEX, 1978-
H.W. Wilson Company. Monthly Except June and December with cumulations in September, November, February and May, and an annual cumulation. (Material from General Science Index is reproduced by permission of the H.W. Wilson Company.)

Arrangement: Alphabetically arranged by subject. Subheadings will be found under broader areas. Under each subject heading and subheading, the entries are alphabetical by title.

Coverage: Astronomy, biology, chemistry, earth science, mathematics, medicine and health, physics, zoology, and others. A list is provided in the "Prefatory Note" of the bound volume and of each issue.

Scope: English language periodicals.

Locating Material: Searcher first determines proper subject area and then scans the material for articles of interest.

Abstracts: Not applicable. Entries provide complete title, author and joint author, but only senior author is given if there are more than two. Bibliographies and illustrative material are noted. Frequent "see" and "see also" references guide the user to alternative terms.

Other Material: Abbreviations used in the entries and for the periodical titles are provided in the preliminary pages of each issue and in the cumulations. There is an author list (or title if there is no author) of citations to book reviews at the end of each issue and the bound volume.

Periodicals Scanned: A list is in the annual cumulation and in each issue.

GOVERNMENT REPORTS ANNOUNCEMENTS AND INDEX, 1946- U.S. Department of Commerce, National Technical Information Service. Semimonthly. Formerly: Government Reports Announcements, Bibliography of Scientific and Industrial Reports, Bibliography of Technical Reports, U.S. Government Research Reports, U.S. Government Research and Development Reports. Absorbed: Technical Translations, 1968.

Arrangement: By subject. Twenty-two subject fields are subdivided into groups. The Committee on Scientific and Technical Information (COSATI) classification scheme is used.

Coverage: Among the subjects included are aeronautics, agriculture, behavioral and social sciences, biological and medical sciences, chemistry, materials, mathematics, military sciences, missile technology, physics, space technology, etc. (See list of subject fields.)

Scope: Government sponsored research and development reports, government analysis sponsored by federal agencies (or their contractors or grantees), federally sponsored translations, and some reports in foreign languages are found.

Locating Material: The entries are arranged alphanumerically by NTIS order number within each section. These numbers, with subject field ordering, and subject group letter were given in the indexes through 1979. Thereafter, page numbers are provided. In the cumulative indexes, volume and issue numbers are also given. A Report Locator List in the back of each issue, arranged alphanumerically by accession number (subject field and group first) was provided until 1975. An edge index to COSATI subject fields is found on the back cover which provides access to subject sections within the journal.

Abstracts: The titles and abstracts are in English. As many as five personal authors and the corporate author are given.

The location of the abstract in Nuclear Science Abstracts, Scientific and Technical Aerospace Reports, Energy Research Abstracts, or Atomindex is indicated. If no abstract is available, this is noted. Contract numbers and the NTIS pricing code are given. A list of descriptors or identifiers was provided through 1979; those with asterisks are access points in the subject index.

Indexes: There are Keyword, Personal Author, Corporate Author, Contract Grant Number, and NTIS Order/Report Number Indexes in each issue. Semiannual and annual cumulations are provided. In the Keyword and Personal Author Indexes, the title, NTIS order number, and page number are given. In the Corporate Author Index, the sponsor's report or series number is also given. The Contract Grant Number Index lists the organization instead of the title. The NTIS Order/Report Number Index gives the report or series number, the title, the NTIS order number, page number and price codes. Before Volume 79, field numbers and group letters were given instead of page numbers. In the cumulative indexes the volume and issue number are given, and price codes are indicated in all indexes. Semiannual and annual cumulations are available. Formerly known as Government-Wide Index to Federal Research and Development Reports, and U.S. Government Research and Development Reports Index.

Other Material: A wealth of explanatory information is provided in the preliminary pages of each issue and the cumulative indexes. An edge index is provided to facilitate the use of this material. Order information is given, order forms and an explanation of the price codes are provided. Abstract newsletters (formerly Weekly Government Abstracts) are available in twenty-seven subject categories. A list of the titles is available and a complete description is provided in the preliminary pages of each issue.

Database: NTIS database has citations from 1964 and is updated twice per month. Abstracts and keywords are provided. Formerly NTISEARCH.

INDEX TO SCIENTIFIC AND TECHNICAL PROCEEDINGS, 1978- Institute for Scientific Information. Monthly with an annual cumulation (semiannual until 1982).

Arrangement: Each issue and the cumulations are in seven sections. Six of these are indexes which refer to the "Proceedings number" used in the Contents of Proceedings section. The numbers which are preceded by the letter "P" are consecutive through each volume.

Coverage: This is a multidisciplinary tool that includes life sciences. agricultural, biological and environmental sciences;

physical and chemical sciences; clinical medicine; engineering, technology, and applied sciences. It indexes published proceedings whether produced as books, reports, or in journals.

Scope: International.

Locating Material: It is necessary to use the indexes to ascertain the proceedings number which will lead the searcher to complete bibliographic information and the titles of all the papers in a given proceedings.

Abstracts: Not applicable. Each entry in the "Contents of Proceedings" section gives complete bibliographic information. The searcher will also find the name of the conference, location, and date of the meeting (if available), up to ten sponsors, citation to the published form (book, journal, etc.) of the proceedings, titles, authors, senior author's address, and starting page numbers of the individual papers. Ordering information is given and the symbol N/A if the proceedings are unavailable, or an indication that they may be obtained without charge. If they may be acquired through ISI's Original Article Text Service (OATS) this is indicated and an accession number is provided. Titles are in English with an abbreviation denoting the language of the original if it is other than English.

Indexes: The category index lists 200 subjects under which proceedings may fall. This enables the searcher to locate proceedings titles in his field of interest. Titles and proceedings number are given. Permuterm Subject Index uses every significant word in titles of papers, conferences, and books and pairs it with every other significant word in a title. Words in the subtitle are also used and are paired separately from the main title. Primary terms are arranged alphabetically with all co-terms listed alphabetically under that. Numbers are found following the Z listings. Only proceedings numbers are given, but this leads the searcher to the "Contents of Proceedings" section. The Sponsor Index lists alphabetically the sponsors of meetings and includes up to ten for each conference. The reference is to proceedings number. The Author/Editor Index lists alphabetically all authors of papers and up to nine editors of each proceedings. Again, only the proceedings number is given. The Meeting Location Index is arranged alphabetically first by country, then by state, then city. The name of the meeting, its date (from 1982) and the proceedings number are provided. The Corporate Index consists of a geographic section and an organization section. In the geographic section, papers are arranged alphabetically by the location (country, state, and city) of the senior author's organization. States of the USA appear first and other countries follow in a separate alphabet. Under each city, the organizations are listed alphabetically. Under each organization the affiliated authors are listed. By each author's name is the proceedings number and the beginning page number of the particular author's paper in the proceedings. The Organization Section is an alphabetical list of the organizations of primary authors which gives the geographic location.

Other Material: Detailed descriptions of how to use the Index are found in the beginning pages of each issue and in the cumulations. Also there are examples from each section which are given on the inside of the front and back covers, and at the beginning of each section. A list of abbreviations used in organizational and geographic names is provided, as is a list for languages and other terms.

Database: SIS/ISTP&B is available through ISI Search Network. This can be accessed from 1978 to the present and includes the proceedings which appear in the printed Index, plus books.

INFORMATION SCIENCE ABSTRACTS, 1966-
Plenum Publishing Corporation. Bimonthly. Formerly: Documentation Abstracts.

Arrangement: There is a subject arrangement with broad topics subdivided into more narrow terms. Within each subject, entries are alphabetically arranged by author (with anonymous articles appearing first).

Coverage: The present classification scheme was initiated with Volume 11. Subjects included are information science documentation, generation, reproduction and distribution of information, its recognition and description, storage, and retrieval. A complete list is found in the Table of Contents for each issue. Journals, conference reports, books and government documents are found.

Scope: International.

Locating Material: Abstracts are numbered consecutively through each volume. A five-digit number is employed, the first two indicating the year and the other three, the consecutive number. These numbers are referred to in the indexes. Frequent cross references lead the user to related items.

Abstracts: The titles and abstracts are in English. The language of the article is indicated if it is other than English and the number of references is noted. Illustrative material is indicated and the address of the senior author is provided. The abstracts range from a single descriptive sentence to a very detailed report and occasionally none appears.

Indexes: There are author and subject indexes in each issue which cumulate annually.

General / 13

Other Material: A list of abbreviations and a list of sources for abstracts are provided. A description of the publication is found in the Introduction.

Periodicals Scanned: A list is provided in the final issue of each volume.

Database: This publication is expected to go online in 1983.

INTERNATIONAL ABSTRACTS IN OPERATION RESEARCH, 1961-
(International Federation of Operational Research Societies)
North-Holland Publishing Company. Four times a year.

Arrangement: By subject within various "orientations" (i.e. Process, Application, Technique, Profession).

Coverage: Among the subjects covered are agriculture and food, biology, ecology, energy, health services, mathematics and mathematical models, computer science, etc. A list is found in the "Digest" of each issue.

Scope: International.

Locating Material: Entries are numbered consecutively from the beginning of the publication. These numbers (serial numbers which are assigned in the order of the occurrence of the entry) are referred to in the indexes. The Digest provides the list of orientations and subjects treated under each. Numerous "see references" in both the Digest and body of the journal lead the user to alternative terms and are occasionally to citations in previous issues.

Abstracts: Titles of papers are in the original language with an English translation if the language is other than English. All authors are given and the address of the senior author is provided. Sources of the abstract are indicated. Titles in the Digest are in English with an abbreviation to indicate the language of the article if other than English.

Indexes: Author and Subject Indexes are found in each issue. These cumulate annually.

Other Material: A description of how to use the journal is provided in each issue. Abbreviation codes are provided for languages and journal titles. Officers and addresses of the Federation are given and related organizations in other countries are noted.

Periodicals Scanned: A list is found in the first issued of each volume.

MASTERS THESES IN THE PURE AND APPLIED SCIENCES ACCEPTED BY COLLEGES AND UNIVERSITIES OF THE UNITED STATES AND CANADA, 1955/56-
Plenum Press. Annual. Previous titles: Master's Theses Accepted by U.S. Colleges and Universities in the Fields of Chemical Engineering, Chemistry, Mechanical Engineering, Metallurgical Engineering, and Physics; Master's Theses and Doctoral Dissertations in the Pure and Applied Sciences Accepted by Colleges and Universities in the United States.

Arrangement: Alphabetically arranged by subject; within each subject, the universities' names are arranged alphabetically and under each university, theses are alphabetical by title.

Coverage: Forty-four disciplines are presently covered which include agriculture, astronomy, chemistry, geology, nuclear science, physics, forestry, transportation engineering, etc. A note on the contents page states: "Mathematics and most life sciences have been excluded from this publication ...to limit the scope of the work. Biochemistry, biophysics and bioengineering are included ... when titles in these areas are reported together with chemistry, physics and engineering...."

Scope: Master's theses from 214 colleges and universities in the United States are found. With Volume 18, the coverage was expanded to include Canadian universities of which twenty-seven are included in Volume 25.

Locating Material: The searcher determines the proper subject and scans the titles under each university. It may be necessary to look in more than one discipline to find a specific title.

Abstracts: Not applicable. The title, date submitted, and the author's name are given for each entry.

Other Material: A list of participating Universities is found in the preliminary pages of each volume. A table of comparative data is given which shows the number of titles reported for each volume.

PANDEX CURRENT INDEX TO SCIENTIFIC AND TECHNICAL LITERATURE, 1967-1972. CCM Information Corporation. Ceased.

Arrangement: Alphabetically arranged by the title of the journal under the major subject category into which it falls. Articles in the particular issues listed are arranged as they appear on the title page of the issue.

Coverage: Selected coverage of science, medicine, technology is offered. Over 2,000 significant journals are fully covered, listing all professional articles and letters in each issue. Included are patents, technical reports, and over 6,000 books.

Scope: International.

Locating Material: Entries are numbered consecutively through each volume. Each article is assigned a six-digit indexing number, 000001, being the first article in the first issue of each year.

Abstracts: Not applicable. Entries give author, title and bibliographic information.

Indexes: An author and subject index is found in each issue.

Other Material: Explanatory material concerning the arrangement, scope and use of the Index is provided in the Introduction.

Periodicals Scanned: A list appears only occasionally.

PROCEEDINGS IN PRINT, 1964-
Proceedings in Print, Inc. Bimonthly.

Arrangement: Alphabetical by the title of the conference. The distinct title is used and not such terms as "Conference on" or "Symposium on."

Coverage: Included are proceedings of conferences, symposia, lecture series, congresses, hearings, seminars, institutes, colloquia, and meetings in all subject areas. Proceedings that are "in preparation" or "in press" are excluded. If it has been determined that the proceedings will never be published or that they are not available for purchase, they are cited and a notation indicating the status is given.

Scope: International.

Locating Material: The entries are consecutively numbered with an accession number. These numbers are referred to in the indexes.

Abstracts: Not applicable. Each entry gives the place and date of the conference and the sponsoring agency. If the title of the proceedings differs from the title of the conference, it is given. Order information is given whenever possible. Complete citations are provided if the proceedings appeared in a journal. Editors' names are provided.

Indexes: An index which includes corporate authors, sponsoring agencies, editors, and subject headings arranged in one

16 / Abstracts and Indexes

alphabet is found in each issue and it cumulates annually.

Other Material: A description of the entries and a brief explanation of the index is found in the preliminary pages of each issue. a list of acronyms and abbreviations for each issue is provided.

REFERATIVNYI ZHURNAL, 1954-
VINITI (All-Union Institute of Scientific and Technical Information, affiliated to Akademiia Nauk, SSSR). Published in series with various beginning dates for the different series. Almost all appear monthly. Some are published in whole or in part as an English translation. Beginning date 1954 represents the first time the publication appeared. Following is a list of the series presently published. Title changes are not indicated.

Astronomiya (Astronomy)
Aviatsionnye I Raketnye Dvigateli (Aircraft and Rocket Engines)
Avtomatika, Telemekhanika I Vychislitel'Naya Tekhnika (Automation, Telemechanics and Computer Technology)
Avtomobil'nye Dorogi (Motor Roads)
Avtomobil'nyi I Gorodskoi Transport (Motor and Municipal Transport)
Biofizika (Biophysics)
Biologiya (Biology)
Dvigateli Vnutrennogo Sgoraniya (Internal Combustion Engines)
Ekonomika Promyshlennosti (Industrial Economics)
Elektronika I EE Primenenie (Electronic Engineering)
Elektrosvyaz' (Electric Communication)
Elektrotekhnika I Elektroenergetika (Electrical and Power Engineering)
Farmakologiya. Khimioterapevticheskie Sredstva. Toksikologiya (Pharmacology. Chemotherapy. Toxicology)
Fizika (Physics)
Fotokinotekhnika (Photography and Cinematography)
Geodeziya I Aeros'emka (Geodesy and Aerial Surveying)
Geofizika (Geophysics)
Geografiya (Geography)
Geologiya (Geology)
Gornoe Delo (Mining)
Gornoe I Neftepromyslovoe Mashinostroenie (Mining and Oil Industry Machines)
Informatika (Information Sciences)
Issledovanie Kosmicheskogo Prostranstva (Space Research)
Khimicheskoe, Neftepererabatyvayushchee I Polimernoe Mashinostroenie (Chemical, Oil-refining and Polymer Machinery)
Khimiya (Chemistry)
Kibernetika (Cybernetics)

General / 17

Kommunal'noe, Bytovoe I Torgovoe Oborudovanie (Municipal, Household and Trading Equipment)
Korroziya I Zashchita ot Korrozii (Corrosion and Protection Against Corrosion)
Kotlostroenie (Boiler-making)
Legkaya Promyshlennost' (Textile Industry)
Lesovedenie I Lesovodstvo (Forestry)
Mashinostroitel'nye Materialy. Konstruktsii I Raschet Detalei Mashin. Gidroprivod (Engineering Materials. Construction and design of machine components. Hydraulic drive)
Matematika (Mathematics)
Meditsinskaya Geografiya (Medical Geography)
Mekanika (Mechanics)
Metallurgiya (Metallurgy)
Metrologiya I Zmeritel'naya Tekhnika (Metrology and Measuring Instruments)
Nasosostroyeniye I Kompressorostroenie. Kholodil'noe Mashinostroenie (Pumps and Compressors. Refrigeration)
Oborudovanie Pishchevoi Promyshlennosti (Food Industry Machinery)
Obshchie Voprosy Patologii (General Problems of Pathology)
Okhrana Prirody I Vosproizvodstvo Prirodnykh Resursov (English not known)
Onkologiya (Oncology)
Organizatsiya I Bezopasanost' Dorozhnogo Dvizheniya.
Organizatsiya Upravleniya (Industrial Management and Organization)
Pochvovedenie I Agrokhimiya (Soil Science and Agricultural Chemistry)
Pozharnaya Okhrana (Protection Against Fire)
Promyshlennyi Transport (Industrial Transport)
Radiatsionaya Biologiya (Radiation Biology)
Radiotekhnika (Radio Engineering)
Raketostroenie (Rocket Engineering)
Rastenievodstvo (Plant Breeding)
Stroitel'nye I Dorozhnye Mashiny (Building and Road Machines)
Svarka (Welding)
Tekhnologiya I Oborudovanie Tsellyulozno-Bumazhnogo I Poligraficheskogo Proizvodstva (Technology and Machinery of Paper-making and Printing)
Tekhnologiya Mashinostroeniya (Mechanical Engineering)
Teploenergetika (Thermal Power Engineering)
Traktory I Sel'skokhozyaistvennye Mashiny I Orudiya (Tractors and Farm Machinery and Equipment)
Truboprovodnyi Transport (Pipelines)
Turbostroenie (Turbine Engineering)
Vodnyi Transport (Water Transport)
Voprosy Tekhnicheskogo Progressa I Organizatsiya Proizvodstva V Mashinostroenii (Problems of Technical Progress and Management in Engineering)

18 / Abstracts and Indexes

>Vozdushnyi Transport (Air Transport)
>Vzaimodeistvie Raznykh Vidov Transport I Konteinernye Perevozki (Coordination of Different Types of Transport. Containers)
>Yadernye Reaktory (Nuclear Reactors)
>Zheleznodorozhnyi Transport (Rail Transport)
>Zhivotnovodstvo I Veterinariya (Animal Husbandry and Veterinary Science)

Arrangement: Arranged by broad subjects subdivided into smaller topics.

Coverage: The journal as a whole covers all disciplines falling in the areas of science and technology, including agriculture and mathematics, but excluding clinical medicine. In addition to periodical literature, monographs, patents, and standards are included.

Scope: International.

Locating Material: Abstracts are in numerical sequence within the broad subject headings in each issue, with the numbering being continuous only within one monthly issue. In issues which are very lengthy, letters are used to indicate sections within the issue. For example, 11K1 refers to the first abstract in Section K of issue Number 11 of a particular year.

Abstracts: Abstracts are in Russian. Titles are in Russian and the language of the article. Oriental characters are transliterated into the Cyrillic alphabet and the Universal Decimal Classification number is often indicated. The abstractor's name is given or an indication of an author abstract. The reference may be followed by letters indicating the type of material that is being abstracted when the material is not a journal article, i.e., whether it is a thesis, map, book, book review, patent, standard, or abstract.

Indexes: While most series have an annual author and subject index, some of the series may have a Geographical Index, Chemical Formula Index and/or a Patent Index. The author indexes are arranged alphabetically within the Cyrillic alphabet and within the Roman alphabet depending on the language in which the article was written. Roman letters are used in the Chemical Formula Index. The Patent Index is made up of three parts: USSR authors; patent holders; country. In the "Biologiya" subject index there is a separate Roman Alphabetical index for the Linnean names of biological groups.

Other Material: A thorough study of this publication has been made by E.J. Copley in A Guide to Referativnyi Zhurnal, 2nd. ed., rev. London, 1972. (National Reference Library of Science and Invention, Occasional Publications.) The writer used this material as well as a study of the journal in compiling the information given here.

General / 19

Periodicals Scanned: A list of the main publications received by a series usually appears in the first issue of the series each year.

SCIENCE ABSTRACTS, 1898-
Institution of Electrical Engineers and the Institute of Electrical and Electronics Engineers, Inc. Published in Series: Ser. A, Physics Abstracts, Semimonthly; Ser. B, Electrical and Electronics Abstracts, Monthly; Ser. C, Computer and Control Abstracts, Monthly. Previous title: Science Abstracts, Physics and Electrical Engineering. First split into two sections (A and B) in 1903. In 1966, Ser. C was added and was originally called Control Abstracts. The description below holds true for all three series.

Arrangement: Arranged by broad subject areas subdivided into more narrow disciplines.

Coverage: The intent of this publication is to cover the whole range of subject matter of the three fields. The back page of each issue carries a classified list of material included. Books, patents, reports and conference proceedings are also covered.

Scope: International.

Locating Material: Entries are numbered consecutively through each volume. These numbers are referred to in the indexes. The subjects under which an abstract falls are given classification numbers and occur in each issue in order by that number. If one wishes to review a given subject, its classification number and the beginning page number is given on the back cover of each issue.

Abstracts: Titles and abstracts are in English. The abstracts are usually brief and concise and occasionally none appears under the author and title citation. Frequently under a subject, one will find a title with a "see reference" to an abstract number in another subject section. If the article is in another language that language is specified and if an English translation is available, a citation for the translation is given. The number of references cited is noted at the end of most abstracts. If an abstract is of a government report, the availability of that report is noted. In most cases, the author's affiliation is given.

Indexes: Author and subject indexes appear in each issue. The author index leads the user to a specific abstract, citing the abstract number. The subject index lists the classification numbers. These indexes cumulate semiannually and are published separately. The semiannual title index also provides the titles of papers and the semiannual subject index gives

20 / Abstracts and Indexes

individual entries. These are arranged under headings in three groups. These groups are: 1) Arabic numbers; 2) English alphabet, Roman numerals, Greek alphabet; and 3) Chemical Formulae. Cumulative author indexes and subject indexes are available for A and B covering the years 1955-59, 1960-64, 1965-68. A combined cumulative author and subject index is available for Ser. C for 1966-68. In addition each issue carries a bibliographic index (to articles containing a significant list of references or a bibliography), a book index, and a conference index (to conference proceedings which are abstracted). A patent and report index was provided until 1974.

Other Material: A description of the journal and instructions for its use are included in each issue. A list of abbreviations and acronyms is provided in the semiannual indexes. A transliteration table for the Cyrillic alphabet is found at the end of the semiannual author index.

Periodicals Scanned: A list is included in the semiannual author index. A supplementary list is provided once a month in Ser. A, and in each issue of Ser. B and C. A list giving additions and amendments is also provided.

Database: Citations from 1969 are available through INSPEC. Abstracts and keywords are available. There are monthly updates for Series B and C (Electrical and Electronics Abstracts and Computer and Control Abstracts). Another file, INSPECT INFORMATION SCIENCE (a special training file) is drawn from the Computer Control section database.

SCIENCE CITATION INDEX, 1961-
Institute for Scientific Information. Quarterly with annual cumulations. A Five-Year Cumulation (1965-1969) is available which includes the Citation Index, Source Index and the Permuterm Subject Index.

Arrangement: There are three parts to this index: the Citation Index, the Source Index, and the Permuterm Subject Index. Entries in the Citation Index are alphabetically arranged by the name of the author being cited. Under his name is the year of publication, an abbreviation for the journal title, the volume, and page. These are arranged chronologically and under each is found the name of authors citing them. To the right of their name is the journal title abbreviation, volume, page, and year for their articles. Titles are not given. The Source Index, which contains Corporate and Personal authors and complete bibliographic information to the citing authors is alphabetically arranged as is the Permuterm Subject Index.

Coverage: This publication is described as an international index to the literature of science, medicine, agriculture, technology, and the behavioral and social sciences. The latter two subjects will be more thoroughly covered in Social Science Citation Index. Coverage constantly expands as the number of journals indexed increases. The technique employed by this index results in a comprehensive, interdisciplinary, elastic index which is always up to date and touches the entire realm of scientific literature.

Scope: International.

Locating Material: Citations are located through the alphabetical arrangement of the references in the Citation Index section. A subject search may be undertaken through the Permuterm Index.

Abstracts: Not applicable. Author, title, bibliographic information, and number of references are given.

Indexes: The Source Index is a complete author index to Science Citation Index (to the citing authors). The Permuterm Index is an alphabetical list of significant words extracted from the titles of all source items processed for the SCI. Each significant word from a title (as a primary term or main entry) is paired with every other significant word in the title (as a co-term or subordinate entry with reference from each such coordination to the first author of an item listed in the Source Index.

Other Material: The "Guide and Journal Lists" is published annually within the set as well as separately. It gives a general introduction to the Index and its components (the Citation Index, the Source Index and the Permuterm Index). Information on arrangement of SCI is found in the preliminary pages of the Citation Index. Special services of ISI are described as are "Conventions Used in the Citation Index." Many of the explanatory paragraphs are repeated in the various volumes. A Statistical Summary, a Bibliography, and reprints of a number of articles on SCI are included.

Periodicals Scanned: A list is published annually in one of the volumes and also in the "Guide and Journal Lists" mentioned above.

Database: Citations are available through SCISEARCH since 1974. There are monthly updates. Keywords and abstracts are not provided.

TECHNICAL BOOK REVIEW INDEX, 1935-
JAAD Publishing Company (Special Libraries Association until December 1976). Monthly except July and August.

Arrangement: Arranged by very broad subject; within a subject the arrangement is alphabetical by author of the book being reviewed.

Coverage: Included are reviews of scientific and technical books within subjects of pure sciences, technology and engineering, life sciences and medicine, general and related fields found in scientific, technical, and trade journals.

Scope: International, though the reviews are found chiefly in U.S. journals and the books are for the most part by American and European publishers.

Locating Material: The searcher looks for the author's name or (in the case of anonymous works), for the title.

Abstracts: Abstracts of reviews are in English. The titles of books being reviewed are in the language of the book. "See references" are made from joint authors to senior author. Complete bibliographic information, including price, is given for the book being reviewed. Length of review in column space and the reviewer's name are given.

Indexes: Annual author (of books being reviewed) index which includes titles of anonymous works is provided.

TRANSLATION REGISTER INDEX, 1967-
National Translations Center. Monthly.

Arrangement: Arranged by subject. Committee on Scientific and Technical Information (COSATI) subject headings are used. The register section announces new accessions to the main NTC collection. The index section includes these and lists other sources of availability in the Directory of Sources which is published with each issue.

Coverage: Subject headings include aeronautics; agriculture; biological and medical sciences; chemistry; earth sciences and oceanography; electronics and electrical engineering; mathematical sciences; mechanical, industrial, civil and marine engineering; navigation; nuclear science and technology; physics; and space technology.

Scope: Translations into English of foreign language material from world literature are found. Material is deposited with the National Translations Center from a variety of sources and information on additional sources is solicited.

Locating Material: A subject approach should be used. The reference number is to be used in ordering copies from the Center. Final digits of the number indicate the COSATI code.

Abstracts: Not applicable. Entries include complete bibliographic data and ordering information.

Indexes: Indexes are found in each issue and semiannual cumulations. They include journal and patent citations in separate sections. Cover to cover translations are not included.

Other Material: Extensive explanatory notes are provided.

WORLD TRANSINDEX, 1978-
International Translations Centre. Monthly. (Merger of World Index of Scientific Translations and List of Translations Notified to the International Translations Centre.)

Arrangement: Arranged by subject. Committee on Scientific and Technical Information (COSATI) subject headings are used.

Coverage: Translations are provided in "all fields of science and technology from East European and Asiatic languages into western languages. Translations from western languages into French, Spanish, and Portuguese are also announced." (Introduction)

Scope: International.

Locating Material: Entries are numbered consecutively through each volume. Indexes refer to these numbers. The first two digits indicate the year.

Abstracts: Not applicable. Entries provide full bibliographic data. The source of articles or the place where the translations may be obtained is given. The original language and languages into which the material is translated are noted.

Indexes: A source index and author index are provided in each issue. These cumulate annually and are published separately.

Other Material: A transliteration table is given in the annual index issue. Brief explanatory material is provided. The text is in English, French, and German.

Periodicals Scanned: The first issue of each volume lists the journals translated cover to cover. A list is found in the annual index issue.

Database: Citations are available through WTI from 1980.

MATHEMATICS, STATISTICS, AND COMPUTER SCIENCE

ACM GUIDE TO COMPUTING LITERATURE see COMPUTING REVIEWS

COMPUTER ABSTRACTS, 1957-
Technical Information Company. Monthly. Formerly: Bibliographical Series (1957-58); Computer Bibliography (1959).

Arrangement: Under each subject, articles in journals are arranged alphabetically by title, followed by U.S. Government Reports. Books are in a separate section.

Coverage: Computer theory, artificial intelligence, pattern recognition, mathematics techniques, programming, system design, data storage and transmission, digital, analogue and hybrid computers, etc. are covered (see Table of Contents). Proceedings, U.S. Government Reports, patents, conference papers, and books are included (see Other Material below).

Scope: International.

Locating Material: Entries are numbered consecutively through each volume. These numbers are referred to in the indexes.

Abstracts: Titles and abstracts are in English. Abstracts are sometimes quite brief, but nevertheless, usually detailed. The number of references is given as is the author's affiliation. If the abstract is of a U.S. Government Report, the AD, CO, or PB serial number is given (a note of availability is found in the preliminary pages of each issue). Frequent "see also" references lead the searcher to related articles or books.

Indexes: Author and subject indexes are found in each issue. These cumulate annually and are published separately. A patent index in each issue was available through Volume 24, 1980. This also had an annual cumulation.

Other Material: Proceedings, U.S. Government Reports, and patents are found among the journal articles. Books are carried in a separate section and full bibliographic information of conference papers and proceedings may be found here. A description of the classification scheme (which indicates in detail the coverage of each section) is issued annually. It usually accompanies the annual index, title page, and table of contents. A note on the availability of articles is in the preliminary pages. A supplement, Computer News, issued separately each month through Volume 24, 1980, is thereafter found as the final section of each issue.

Periodicals Scanned: A list is found in the annual index.

COMPUTER AND CONTROL ABSTRACTS see SCIENCE ABSTRACTS

COMPUTER AND INFORMATION SYSTEMS ABSTRACT JOURNAL, 1962- Cambridge Scientific Abstracts. Monthly. Formerly: Information and Processing Journal, 1962-1978.

Arrangement: Arranged by subject. Subject headings are assigned letters, not numbers. These are listed in the Table of Contents for Volume 27, number 1, with detailed subdivisions. Brief subdivisions of the main headings are given in later issues.

Coverage: Books, journal articles, conference proceedings, dissertations, government reports, and patents are included. Computer software, applications, mathematics, and electronics are covered. Under Applications, the searcher can expect to find artificial intelligence, information retrieval, electronics and communication; under Computer Mathematics such subjects as calculus, abstract mathematics, and probability and statistics are covered; a general section includes economic considerations, legal problems, security, and personnel.

Scope: International.

Locating Material: Entries are numbered consecutively through each volume. The number consists of seven digits (the first two indicate the abstract's place in the sequence). These numbers begin with 0001 with each volume.

Abstracts: Titles and abstracts are in English. If the language is other than English, the title may be given in the original and summaries in other languages are indicated after the title. Up to ten authors' names are given and the address of the senior author is provided. The abstracts give a good indication of the content of the paper.

Indexes: Acronym (nomenclature), subject, and author indexes are in each issue. These cumulate annually. A source index is found in issue number one of each volume.

Other Material: A list of abbreviations and a brief guide to using the journal is provided in the preliminary pages.

Periodicals Scanned: The Source Index (number one of each volume) provides the list of journals examined. This is updated semi-annually.

Database: ELCOM (Electronics and Computers) provides entries from Computer and Information Systems Abstract Journal from 1978. Updates are provided every two months and abstracts are available but not keywords. (Electronics and Communications Abstracts Journal is also in this database.)

COMPUTER PROGRAM ABSTRACTS, 1969-1981.
National Aeronautics and Space Administration; for sale by the Superintendent of Documents. Ceased.

Arrangement: Arranged by broad subjects subdivided into more specific categories.

Coverage: The abstracts are of "documented computer programs developed by the National Aeronautics and Space Administration, NASA Contractors, and other Federal Agencies...." (Introduction). Among the subjects covered are aeronautics, astronautics, engineering, geosciences, life sciences, mathematics, and space science.

Scope: Material developed by United States agencies and institutions.

Locating Material: Each entry is assigned a sequential accession number (the letter "M," two digits indicating the year, and the sequential number). These numbers are referred to in the indexes.

Abstracts: The corporate source, personal author, title, program language, a program number, machine requirements, and price are given. The abstracts are usually very thorough and fairly long. "See references" at the beginning of subject sections lead the reader to alternate terms.

Indexes: A subject index, an originating source index, program number/accession number index, and an equipment requirements index are found in each issue. A cumulative issue was published in 1971 which covered July 15, 1969-July 15, 1971. This superseded the previously published issues and indexes to them.

Other Material: An explanation of the entries is given in the preliminary pages of each issue. The Table of Contents lists the material included in each subject section and subdivision.

COMPUTING REVIEWS, 1960-
Association for Computing Machinery. Monthly.

Arrangement: Arranged by broad subjects subdivided into narrower topics. With Volume 22, the issues were divided into sections: Books and Proceedings, and Literature other than Books (i.e., journal articles, etc.). Each of these sections has the subject/category arrangements.

Coverage: The stated aim is to provide "...critical information about all current publications in any area of the computing sciences..." The categories include application (natural sciences, engineering, humanities), software (programming languages, supervisory systems, evaluation, etc.), hardware (systems, components, and circuits, etc.), mathematics of computation, and analog computers.

Scope: International.

Locating Material: The entries are numbered consecutively from Volume 1. These numbers are referred to in the indexes.

Abstracts: These are reviews rather than abstracts in most cases and range in length from a short paragraph to more than half a page. The titles are in English. The senior author's affiliation is given and the reviews are signed. Frequent "see also" references at the beginning of subject categories lead the researcher to related items.

Indexes: An author index is found in each issue which cumulates annually and a cumulative index covering 1960-1976 is available. Included are an author index, a permuted keyword index, a classification, reviewer, and source index. The ACM Guide to Computing Literature provides an enhanced index to Computing Reviews. It provides a Bibliographic Listing, an Author Listing, a Keyword Index, CR Category Index (complete with an Index to the Categories), Computing Reviews Reviewer Index, and a Source Index. There are a great many more items indexed than those found in Computing Reviews. The entries in the indexes are identified by a five-digit number referring to a CR entry or a four digit number indicating an unreviewed publication. Bibliographic information for these is found in the bibliographic listing. The four-digit numbers lead the searcher to the correct entry.

Other Material: Each issue carries a quotation on the front cover. The source of the quotation is given on the inside back cover. A list of editorial symbols is provided. The Classification System for Computing Reviews is found at the end of most issues.

Periodicals Scanned: A list of periodicals and of books received is occasionally included in issues.

Mathematics, Statistics, and Computer Science / 29

CURRENT MATHEMATICAL PUBLICATIONS see MATHEMATICAL REVIEWS

MATHEMATICAL REVIEWS, 1940-
The American Mathematical Society. Monthly. Two volumes a year until 1980 when the volume numbering changed. With that year each issue was designated by an 80 and a letter--A for January through K for November (skip L) for December. (In the old system, 1980 would have been Vol. 59-60.)

Arrangement: Arranged by subjects.

Coverage: The field of mathematics and mathematical applications is given excellent coverage. The user should refer to the Table of Contents for a list of subjects.

Scope: International.

Locating Material: Entries are numbered consecutively through each volume (two volumes a year).

Abstracts: The abstract (or review) is in English. Titles are usually in the language of the original and in English, but sometimes in English only and sometimes in the language of the article only. Translations are noted. Journal titles are usually in the language of origin. Abstracts range in length and detail from one line to a column or more. Occasionally none will appear. If the review is from another reviewing or abstracting journal, the source is identified by initials following the reviewer's name (a list is on the verso of the front cover). A star by an entry indicates it is a book or other nonserial publication being reviewed as a whole. "See also" references send the searcher to other entry numbers relating to his subject (found at the beginning of a subject section).

Indexes: An author index is found in each issue and from Volume 80, a Key Index is provided. There is an annual author and subject index. Each volume has a separate index issue (appearing in June and December) which has an author index and a key index. A list of subject classifications is provided from time to time (see 1980 Annual Index). Cumulative author indexes are available for 1940-59, 1960-64, and 1965-72, and a cumulative author and subject index for 1973-79.

Other Material: The index issues also give a table for transliteration of the Cyrillic alphabet, errata, and addenda. Explanatory material appears in each issue and in the indexes. Current Mathematical Publications, also published by the American Mathematical Society, is produced in the editorial offices of Mathematical Reviews. Each issue contains a list of the material received by Mathematical Reviews during a specific length of time. A subject index classifying the material comprises most of an issue. There is also a list of the journals received, and the tables of contents of those journals when they are available. A star (*) by a title indicates that it is a book.

Periodicals Scanned: A list in the index issues gives the abbreviations, full title and place of publication for each journal title. Additions and changes are found in each issue.

Database: Entries are available through MATHFILE from 1973. The updates are monthly.

NEW LITERATURE ON AUTOMATION, 1960-
Studiecentrum NOVI (Amsterdam, The Netherlands). Monthly except July/August.

Arrangement: Arranged by subject.

Coverage: Topics covered vary with each issue but these include computer industry, personnel and systems, database management and data transmission, artificial intelligence, programming languages, information storage and retrieval, software and hardware, medical applications, etc. The contents of each issue provide a list for that issue.

Scope: International.

Locating Material: Abstracts are numbered consecutively through each volume. There are two digits indicating the year and a slash before the consecutive number.

Abstracts: Titles and abstracts may be in English, German, or French (an edition for Dutch users is also available). An indication of the form of material the entry is from is given (i.e., a book, a journal article, etc.). Up to three authors are given. A list of subjects treated (keywords) in each entry is provided at the beginning of the entry.

Indexes: An annual author index and keyword subject index are provided.

Other Material: A brief "Digest" on various topics is found in each issue. A list of subject categories is provided separately.

STATISTICAL THEORY AND METHOD ABSTRACTS, 1959-
The International Statistical Institute by Longman Group, Ltd. Quarterly. Formerly: International Journal of Abstracts; Statistical Theory and Method.

Mathematics, Statistics, and Computer Science / 31

Arrangement: Arranged by subject. Broad areas with subtopics are found. Classification numbers are provided (scheme is in the preliminary pages of each issue). Formerly the pages were color-keyed for different broad areas with the color noted in the classification scheme.

Coverage: Mathematical statistics and probability theory including games and decision theory, frequency and sampling distributions, estimation, hypothesis testing, variance analysis, sampling design and experimental design, stochastic processes (see Classification Scheme). There are twelve subject divisions with up to twelve subdivisions.

Scope: International.

Locating Material: Abstracts are numbered consecutively through each volume. The numbers consist of a volume number, a slash and the abstract number. In the author index, the abstract number and a classification number is given.

Abstracts: Titles and abstracts are in English. The language of the article is indicated and the senior author's address is provided if available and if not, that of one of the other authors is given. The number of tables and references in the text is given.

Indexes: An author index is in each issue which lists all authors. An annual index is provided which gives a compilation of the separate indexes in each issue. Explanatory material is provided.

Other Material: An analysis of the secondary classification which shows its relationship to the primary classification is provided. A list of new statistical tables and statistical algorithms was provided for earlier volumes but discontinued about 1975. An explanation of the abbreviations used is given in the introductory material. A description of the entries, how they are arranged, and their components is given.

Periodicals Scanned: A list of "New Journals Represented" is found in each issue and a list of abbreviations of names of journals is provided in the annual issue.

ZENTRALBLATT FÜR MATHEMATIC UND IHRE GRENZGEBIETE, 1931-
Springer-Verlag. Twenty volumes a year. Until Volume 218, issued in two sections. Number 1-2; with Volume 221, no longer has notation "heft 1/2." Suspended between 1944 and 1948.

Arrangement: Arranged by subject.

Coverage: All fields of pure and applied mathematics, including theoretical mechanics, history, logic, set theory, algebraic geometry, topological groups, special functions, approximation and expansions, probability theory, etc. are covered (see Table of Contents). Books and other nonserial publications are included among the journal entries. Articles from proceedings volumes usually are reviewed separately.

Scope: International.

Locating Material: Subject sections are numbered from 00 to 94. A list in each issue gives page locations of the beginning of the subject sections. Many of the sections are further subdivided according to the MOS subject classification scheme of the American Mathematical Society. The identifying number consists of five digits. The first two digits refer to the subject sections number. The rest is sequential from 001 through each section. "See references" at the beginning of each subject section lead the user to related entries.

Abstracts: If the article is written in English, German, French, or Italian, the title will be in the language of the article. Otherwise it is translated, usually into English or German and the language of the original is indicated. Summaries in other languages are usually noted. Abstracts are also written in one of these four languages with Italian occurring only rarely. Source of the abstract is given (abstractor's name, abbreviations for other abstract journals, or author abstract). Abstracts are usually detailed, some running to more than a page.

Indexes: An author index is in each volume and from Volume 239, subject indexes are also found. Every tenth volume (e.g., Volume 220, 230, etc.) contains cumulative indexes to personal authors, a key index (previously called anonyma) which lists congress proceedings, other collections of articles, or works with no personal authors (listed by a keyword), biographical works, and a list of corrections. From Volume 250 there will be a cumulative subject index in every 10th volume. All references are to volume number and abstract numbers except the errata list which is to page number. The first ten-volume index is Volume 110. Every fiftieth volume is a cumulative index for the preceding 49 volumes. The first 50 volume index is Volume 300.

Other Material: Transliteration table for the Cyrillic alphabet is provided. A description of the current index system and for Volumes 1-100 is found in the preliminary pages.

Periodicals Scanned: A list is in every 10th volume with the cumulated indexes. It gives abbreviations, full title, the publisher and place of publication.

Database: Entries are available from 1978 through INKA-MATH. Abstracts and keywords are provided.

ASTRONOMY

ASTRONOMISCHER JAHRESBERICHT, 1899-1968.
Walter de Gruyter and Company. Ceased. Superseded by
Astronomy and Astrophysics Abstracts.

Arrangement: Arranged by 25 broad subject headings indicated by
Roman numerals, and subdivided into smaller areas indicated
by Arabic numerals.

Coverage: The subject matter is covered broadly and includes
general information such as bibliographies, histories, biographies, planetaria, societies and institutions, as well as
mathematical, physical, and other related information. (See
Contents for a list of subject headings.)

Scope: International.

Locating Material: The citations in each subsection are numbered
separately. The first digit(s) of the number indicate the
section and are followed by the sequential number of the item.
Two or more digits are always used for the item number.
For example: 101 is the first item in subsection 1; 1912 is
the twelfth item in subsection 19; 7603 is the third item in
subsection 76.

Abstracts: Very brief abstracts are given the first time the article
is cited. An article may be cited under several subjects, in
which case no abstract is given in the later citings but a
reference is given to the earlier abstract. If the abstract
has been taken from another abstracting tool, the abstract is
not repeated, but a reference is given to the original source.
The abstractor's initials are given with the citation.

Indexes: Author and brief subject indexes are provided.

Other Material: A list of abbreviations and a transliteration of the
Russian alphabet is provided. A list of abstractors is found
in the Foreword.

Periodicals Scanned: A list appears in Section I, 1, Bibliographie,
in each volume. The full title of the periodical is given,
with a list of volumes or numbers indexed.

ASTRONOMY AND ASTROPHYSICS ABSTRACTS, 1969-
   Springer-Verlag. Semiannual; two volumes cover a calendar
   year. Supersedes: Astronomischer Jahresbericht, 1899-
   1968.

Arrangement: Divided into 108 numbered subject categories (the
   numbers are 001-162 as some numbers are skipped) representing in logical sequence, the subject covered. Some subject categories are subdivided into smaller groups, with the manner of the subdivision being described in the Introduction.

Coverage: This is a comprehensive documentation of literature in
   all areas of astronomy and astrophysics. Popular articles are listed but not abstracted and border areas, according to their relevance to astronomical research, are represented but not always abstracted. Short news notes, books on astronomy and astrophysics, and related disciplines are indexed. There are some references to obituaries, personal notes, proceedings of colloquia, congresses, meetings, symposia, and book reviews. (See Contents for outline of subjects covered.)

Scope: International.

Locating Material: Each item is allocated a serial number which
   consists of three parts, a period dividing each part. The first represents the volume in which the reference occurs. The second group of numbers represents the subject category and the third group represents the serial number of the item within the category. For example: 06.031.076--the 06 represents volume 6; the 031 represents a subject category (in this case, "Optics, Methods of Observation and Reduction") and 076 indicates that this is the 76th citation under that subject. The first group of numbers (06) appears only at the top of each page and not with each item. References in the index refer only to the second and third group of numbers (031.076).

Abstracts: The titles of papers are given in the language of their
   authors when possible with English translations provided if available. Titles of Russian papers are given in English. The author's abstract is generally used, otherwise the abstractor's name is given. Entries from other abstracting tools which are not included are added. When a paper is listed under more than one subject, its abstract is given in only one category and cross referenced from other categories. Abstracts are in English, French, or German, but most appear to be in English. The abstracts are brief but full bibliographical information is given.

Indexes: There are author and subject indexes in each volume.
   The introduction states "...the magnetic tapes containing the index information will be used to produce separate index volumes (author and subject) at intervals of five years."

Other Material: The Introduction contains explanatory material concerning the abstracts and the indexes. There is also a transliteration of the Russian abphabet, a discussion of the sources of information and of the classification scheme. A list of abbreviations and acronyms is also provided. Further, the introduction in each issue contains a list of six important bibliographies which cover the literature of astronomy from 480 BC to date. They are:

1. J.J. de Lalande, Bibliographie Astronomique, Paris, 1803. This work covers the time from 480 BC to the year 1803.
2. J.S. Houzeau, A. Lancaster, Bibliographie générale de l'astronomie, Volume I (in two parts). Bruxelles, 1882, 1887; Volume II, Bruxelles, 1889. The complete title of Volume II is Bibliographie générale de l'astronomie ou catalogue méthodique des ouvrages, des mémoires et des observations astronomiques publiés depuis l'origine de l'imprimerie jusqu'à 1880. A new edition of these volumes was prepared by D.W. Dewhirst in 1964.
3. Bibliography of Astronomy, 1881-1898. The literature of this period was recorded on standard slips by the Observatoire Royale de Belgique. From the material a microfilm version was produced by University Microfilms Limited, Tylers Green, High Wycombe, Buckinghamshire, England, in 1970.
4. Astronomischer Jahresbericht, 1899 gegründet von Walter Wislicenus, herausgegeben vom Astronomischen Rechen-Institut im Heidelberg. Verlag W. de Gruyter, Berlin. For the period from 1899 to 1968 sixty-eight volumes were published, each of which in general, covers the literature of one year.
5. Bulletin Signalétique - Section Astronomie, Astrophysique, Physique du Globe. Published by Centre de Documentation du Centre National de la Recherche Scientifique, Paris. This publication is a continuation of Bibliographie Mensuelle de l'Astronomie founded in 1933 by the Société Astronomique de France. The publication is continued.
6. Referativnyi Zhurnal. Founded in 1953 and published by Vsesoyuznyj Institut Nauchnoj i Tekhnicheskoj Informatsii, Akademiya Nauk, Moscow. The publication is continued.

Periodicals Scanned: A list of the publications abstracted or indexed appears in sections 001, Periodicals, and 008, Observations, Institutes. Titles of periodicals are given in their original languages; Russian titles are transliterated. The sections 008, Records, and 150, Periodicals, include publication series of observatories and astronomical institutes which are not included in section 001. The place and publisher of each serial are given.

Database: It is planned that citations will be available through INKA-ASTRO.

METEOROLOGICAL AND GEOASTROPHYSICAL ABSTRACTS, 1950-
American Meteorological Society. Monthly. Formerly:
Meteorological Abstracts and Bibliography, 1950-1960.

Arrangement: Arranged by broad subjects subdivided into narrower topics.

Coverage: The areas of meteorology, climatology, aeronomy, planetary atmospheres, and solar-terrestrial relations are thoroughly reported. This also covers substantially the fields of hydrology, oceanography, glaciology, cosmic rays, radio-astronomy, air pollution, and sea-air interaction as they relate to the atmospheric sciences. Meteorological applications to or from aeronautics, agriculture, astronautics, biology, forestry, industry, engineering, public health, and radio propagation are also included. In addition to periodical literature, the publication contains irregular serials, reports, conferences, compendiums, textbooks, and daily reports. Also covered are theses, yearbooks, publications, reviews, data, bibliographies, translations, glossaries, dictionaries, and patents.

Scope: International.

Locating Material: Entries are numbered consecutively through each issue. The abstract number also includes the number of the volume and of the issue. Thus, 23-10-9 refers to the ninth abstract in number ten of Volume 23.

Abstracts: The abstracts are in English. Titles are in English and in the language of the article unless otherwise stated. Cyrillic titles are transliterated into the Roman alphabet and available summaries in translation are noted. The number of references, illustrative material, and appendices, etc., are indicated. The Universal Decimal Classification number and the subject headings under which the citation will fall are given. For most entries, a symbol for at least one library where the material may be found is provided. "See also" references are scattered throughout the abstracts following the subject headings. Titles of periodicals are not, as a rule, abbreviated and place of publication for journals is given.

Indexes: Author, subject, and geographical indexes are found in each issue. There is a cumulative author, subject and geographical index for each volume. Cumulative author and geographic area indexes for volumes 1-10, 1950-1960 are available.

Other Material: Definitive bibliographies on many subjects (such as Medical Meteorology, Climate of the Arctic, Thunderstorms, etc.) appeared at intervals as Part II of the journal through Volume 15, no. 12, 1964. An alphabetical list of these bibliographies is printed in each issue and in the annual index. (American Meteorological Society. Cumulated Bibliography and Index to Meteorological and Geoastrophysical Abstracts, 1950-1969; classified subject and author arrangements. Boston: G.K. Hall, 1972, 9 Volumes, is a cumulated index to this publication.)

Database: Citations are available through MGA from 1972. Abstracts are available for records of 1972-73 and from 1976 to date.

CHEMISTRY AND PHYSICS

ACOUSTICS ABSTRACTS, 1967-
Multi-Science Publishing Company Ltd. Bimonthly.

Arrangement: Arranged by subject; some broad areas have subdivisions. The publication is in two sections, A and B, but
the same subjects are covered in each.

Coverage: Fundamental, solid state, liquid state, gaseous state
acoustics; measurement and techniques; materials and devices;
ultrasonic applications; underwater acoustics; audio frequencies;
recording and reproduction; physiological, psychological, and
bioacoustics; noise; architectural acoustics; and vibrations
and shock are included.

Scope: International.

Locating Material: Abstracts are numbered consecutively through
each volume and separately for each section (a "B" is added
for entries from that section).

Abstracts: Titles, abstracts, and journal titles are in English.
Senior author's affiliation is given and illustrative material
and the number of references are noted.

Indexes: An annual author and subject index (which includes both
sections) is provided.

AMINO-ACID, PEPTIDE AND PROTEIN ABSTRACTS see BIOCHEM-
ISTRY ABSTRACTS

ANALYTICAL ABSTRACTS, 1954-
The Royal Society of Chemistry. Monthly (two volumes per
year). Supersedes British Abstracts, Ser. C., Analysis
and Apparatus (1944-1953).

Arrangement: Broad subject areas are subdivided into narrower
topics.

Coverage: Deals with all branches of analytical chemistry; general
analytical chemistry, inorganic chemistry, organic chemistry,
food, agriculture, air, water, effluents, techniques, apparatus,
chromatography, ion exchange, electrophoresis, optical,

electrochemistry, radiochemistry, thermal analysis, and particle size analysis. Monographs, reports, conference proceedings, and symposia are included.

Scope: International.

Locating Material: Entries are numbered consecutively through each volume.

Abstracts: Titles and abstracts are in English. The language of the original article is specified. Abstractor's name and location of the senior author is given. Each subject area of abstracts is followed by a list of cross references.

Indexes: Author and subject indexes are provided for each volume in a separate issue. Patents are so indicated (by the letter P) as are conference papers (by the letter C) and books (by the letter B).

Other Material: A list of abbreviations and units of measure are included in each issue and a list of recent books appears periodically. These are in the Volume Index along with an explanation titled "Introduction to the Index" and a description of the journal sections, scope, and entries. A list of the abstractors is provided.

Periodicals Scanned: A list appears in the beginning of the index issue and contains the names of journals (and their abbreviations) from which papers have been abstracted regularly in the volume.

Database: Analytical Abstracts is expected to have an online version in 1984.

BIOCHEMISTRY ABSTRACTS, Part 1: BIOLOGICAL MEMBRANES, 1973-   ; Part 2: NUCLEIC ACIDS, 1971-   ; Part 3: AMINO-ACIDS, PEPTIDES AND PROTEINS, 1972-   . (The three parts were formerly issued separately under their own titles.)
Cambridge Scientific Abstracts. Monthly.

Arrangement: Arranged by subject.

Coverage: Subjects included in Part 1 (Biological Membranes) are chemical components; physical properties, model systems and theoretical aspects of membrane functions; isolation techniques; morphology; transport and related phenomena. Part 2 (Nucleic Acids) includes purines, pyrimidines and analogs; nucleosides and analogs; nucleotides, nucleoside di- and triphosphates and analogs; olegonucleotides, polynucleotides and analogs; transfer RNA; protein biosynthesis; TNA; DNA; enzymes; immunological aspects; and protein-nucleic acid association. Part 3 (Amino-Acids, Peptides, and Proteins) covers amino acids; purification and preliminary characterization of peptides and proteins; conformation and solution studies of peptides and proteins; X-ray, electron microscope and neutron diffraction studies; theoretical aspects of protein structure; peptide synthesis; enzyme essay; mechanism of enzyme action; immobilized enzymes; peptide and protein hormones and related substances; insect and reptile toxins and related substances; peptides with structural features not found in proteins; and evolutionary aspects.

Scope: International.

Locating Material: Entries are numbered consecutively through each volume. There are three parts to these numbers: first a number for the abstract, then a code letter identifying the journal, and finally the volume number.

Abstracts: Titles are in English and in the language of the article. The language and summaries in other languages are indicated. All authors are named and the senior author's affiliation is given.

Indexes: Author and subject indexes are provided in each issue. These cumulate annually.

Other Material: There is a section explaining how to use the journal, some descriptive information is provided, and a list of abbreviations is given.

Periodicals Scanned: A list of periodicals is available on request. A separate list covering Cambridge Scientific Abstracts' life science abstract journals is available for sale.

Database: Citations are available from 1978 through IRL LIFE SCIENCES COLLECTION. Keywords and abstracts are provided.

BRITISH ABSTRACTS see ANALYTICAL ABSTRACTS

CHEMICAL ABSTRACTS, 1907-
American Chemical Society, Chemical Abstracts Service.
Weekly with Volume Indexes in June and December.

Arrangement: Arranged by subject; divided into eighty abstract sections. Within subject sections, journal articles, proceedings, edited collections, technical reports, and dissertation abstracts are first, new book announcements second, and patents third. These groups are divided by dashes placed across the column. Two consecutive weekly issues are required to obtain all 80 sections.

Coverage: An extremely wide range of material concerning chemistry and chemical engineering is included. Broad subjects are: biochemistry, organic chemistry, macromolecular chemistry, applied chemistry, and chemical engineering, physical, and analytical chemistry. (See the Table of Contents for the eighty subject sections under these headings.)

Scope: International.

Locating Material: Each reference is designated by an abstract number. These abstract numbers run continuously through a six-month volume. A letter following the number is a computer-generated check letter by which a reference is computer validated (in use since Volume 66, 1967). These should not be confused with column fraction designations used earlier. These numbers and letters are referred to in the Issue, Volume, and Collective Indexes.

Abstracts: Abstract headings include document title, complete bibliographic citation, and names and affiliation of authors, and patent assignees. Journal titles are abbreviated and are found unabbreviated in Chemical Abstracts Service Source Index (CASSI). Abstracts and titles are in English and the language of the article is designated. All authors' names are given unless there are more than ten. The address where the work was done or where correspondence regarding the article should be sent is given. In proceedings of a meeting, the date of the meeting as well as the publication date (if different) is supplied. For books, complete bibliographic citation including the number of pages and the price is given. For patents, the title may be augmented or reworded. The assignees' as well as the inventors' names and the patent number with an abbreviation indicating the country granting the patent are given. Cross references at the end of each section lead the user to related abstracts.

Indexes: Issue Indexes include Keyword Subject, Numerical Patent Indexes, Patent Concordance, and Author Index. Semiannual (Volume indexes) include Author, Chemical Substance, General Subject, Formula, Index of Ring Systems, Patent, and Patent Concordance. Decennial Indexes were published from 1907-1956, five-year Collective Indexes starting in 1957 are provided. Current Collective Indexes include Chemical Substance, General Subject, Author, Numerical Patent, Index of Ring Systems, and Formula Indexes. Also included are a Patent Concordance and an Index Guide. Special Collective Indexes include a 27-Year Collective Formula Index to Chemical Abstracts (1920-1946) and a 10-Year Numerical Patent Index to Chemical Abstracts (1937-1946).

Other Material: Chemical Abstracts Service (CAS) makes every effort to facilitate the use of the abstracts through the provision of explanatory material from which much of this discussion is taken. This is found in the preliminary pages of the weekly issues, the indexes and CAS Source Index (CASSI). Also provided are a list of abbreviations and symbols used in ACS publications, illustrative material to demonstrate entries, a list of language abbreviations, a list of abstractors, and suggestions for procurement of copies of documents. Information regarding other CAS abstracting and indexing services may be obtained from Chemical Abstracts Service.

Periodicals Scanned: Chemical Abstracts Service Source Index replaces previous editions of ACCESS, List of Periodicals Abstracted by Chemical Abstracts, (1908-1960) and CA-List of Periodicals (1961-1967). There is a cumulation covering 1907-1979 and Quarterly Supplements. The fourth quarterly supplement of a year cumulates and replaces the other three. It contains bibliographic data and library file location information. All journals and nonserial literature covered by CA are entered. Information about publications added to the coverage and changes in publication titles is found in CASSI and in a separate section following the Author Index in even numbered issues of CA.

Database: Entries are available from 1967 through CA SEARCH. There are updates twice per month; keywords are provided but abstracts are not.

CHEMISCHES ZENTRALBLATT, 1830-1969.
Deutsche Chemische Gesellschaft. Ceased. Formerly: Pharmaceutisches Centralblatt, 1830-1949; Chemisches-Pharmaceutisches Centralblatt, 1850-1855; Chemisches Central blatt, 1856-1906. In 1919 took over the abstract section of Zeitschrift für Angewandte Chemie. Between 1947 and 1949, appeared in two editions, one published in East Berlin, the other in West Berlin using the same volume numbers but with different content; there was also an American edition.

Arrangement: Arranged by subject. There are broad areas with subtopics.

Coverage: General organic and inorganic chemistry, mineralogical and geological chemistry, biochemistry, laboratory analysis and applied chemistry are among the subjects (see Table of Contents). Patents and bibliographies are included.

Scope: International.

Locating Material: No numbering system is used. Material is gathered under the subject headings (which are assigned letters) and subheadings (which have the subject letter and either a number, a small letter or a Roman numeral). An author index on the inside front and back cover leads one to the proper page. The author index indicates corporate entries.

Abstracts: Titles and abstracts are in German. All authors are given, and the source of the article is at the end of the abstract. "See references" refer the user to related articles. Abstractor's name and senior author's affiliation are give.

Indexes: An author index is in each issue. A patent index is given at the end of the year. A subject index, index of formulas or organic compounds (listed according to M.M. Richter's system) for the year are available. Bibliographies are listed separately at the end of the Author Index with page references. Some cumulative indexes are published: Volumes 41-52; Quinquennial from Volume 68- (1897-19).

CHEMO-RECEPTION ABSTRACTS, 1973-
Cambridge Scientific Abstracts. Quarterly.

Arrangement: Arranged by subject. Broad areas are subdivided into more narrow topics.

Coverage: Broad areas include peripheral and central sensory mechanisms, psychophy, human physiology and pathology, neuroanatomy-histology, histochemistry, chemotaxis, animal behavior, and chemical communication including pheromones, chemistry of odorous materials, chemistry of sapid materials, odor control, chemo-sensory aspects of food, etc. (see Contents for a complete list).

Scope: International.

Locating Material: Entries are in numerical sequence through the volume. First is a consecutive number, then the letter code identifying the journal, then the volume number.

Abstracts: Abstracts are in English. Titles are in English and for European languages they are also given in the original. The language of the article and of summaries in translation are indicated. Up to ten authors' names are given and the address of the first author is provided. The author's abstract is used when possible. The abstracts attempt to be informative and range in length from 150 to 200 words. Indicative abstracts are provided for extremely long review articles.

Indexes: Author and subject indexes in each issue cumulate annually.

Other Material: Book Notices and Notification of Proceedings are found in each issue. A key to abbreviations used in the abstracts and to language abbreviations is provided in each issue. Explanatory material is in the first issue of the volume.

Periodicals Scanned: A complete list of periodicals searched by Cambridge Scientific Abstracts for the life sciences abstract journals is available for sale.

Database: Citations are available through IRL LIFE SCIENCES COLLECTION from 1978. There are monthly updates with abstracts and keywords available.

CURRENT ABSTRACTS OF CHEMISTRY AND INDEX CHEMICUS, 1960-
Institute for Scientific Information. Weekly. Formerly: Index Chemicus, 1960-1969. (From 1970, Index issues have the title Index Chemicus).

Arrangement: The abstracts are organized by journal and are given a sequential number from the beginning in 1960.

Coverage: All new compounds, new reactions, and new syntheses reported in the source journals are included. These report the most significant worldwide research in the fields of organic, pharmaceutical, medical, and biological chemistry.

Scope: International.

Locating Material: Entries are consecutively numbered through the volume.

Abstracts: Titles and abstracts are in English. The language of the original article is noted. Authors' names and addresses are given and ISI (Institute for Scientific Information) accession number is provided. Abstracts are in a narrative summary provided by the author and are graphic; flow-diagrams are used extensively. Use-Profile and technique data symbols alert the user to activities for which compounds were tested and to analytic procedures used by the investigator.

Indexes: Index Chemicus is incorporated into the weekly issues and cumulates quarterly and annually. It consists of five parts: molecular formula, subject, author, biological activity, corporate, and an alert to labeled compounds. The cumulations include a Rotoform Index of molecular formulas. Molecular formulas are identified by both an abstract number and a compound designation. The unique abstract number refers to a particular article in CAC and IC and the compound designation matches an asterisk or an underscored number or letter within the article. An asterisk indicates an old product of a new reaction or a new synthesis. Underscored letters or numbers indicate new compounds. All compound numbers appear as Arabic numbers in the index. Only in articles where the author uses both Roman and Arabic numbers for diagram designation will the Roman number be translated with the prefix R. In cases where Roman numerals are the majority, an A may prefix the Arabic numbers only. The Rotaform Index contains all molecular formulas included in the molecular formula index. However, those compounds containing only carbon, hydrogen, nitrogen, and oxygen have been omitted. In the Rotaform Index, the order of the elements in the molecular formula is "rotated" so that each element appears as a heading in the index. For each different molecular formula indexed there is a separate entry.

Other Material: The first page of each issue of CAC is a table of contents which indicates the source journals covered in that issue. There is also a guide to the labeled compounds, new reactions and new syntheses highlighted in the issue. Brief explanatory material is found in the preliminary pages.

Periodicals Scanned: A list is found in each issue and in Index Chemicus.

CURRENT CHEMICAL REACTIONS, 1979-
Institute for Scientific Information. Monthly.

Arrangement: Entries are grouped by the source journal.

Coverage: The intent is to "...provide a guide to new and newly modified reactions and syntheses reported in current journal literature." (Introductory remarks.)

Scope: International.

Locating Material: Entries are numbered consecutively through the volume.

Abstracts: "Notes for the User" states "...summaries and flowcharts of new and newly modified reactions and syntheses reported... (and) other reaction data are provided...."

Indexes: Four indexes are found in each monthly issue: author, journal, corporate, and permuted subject. These cumulate annually.

Other Material: Brief explanatory material is provided.

Periodicals Scanned: A list of titles of books and journals covered appears occasionally.

CURRENT PHYSICS INDEX, 1975-
American Institute of Physics. Quarterly.

Arrangement: Arranged by subject with subdivisions for subtopics.

Coverage: The physics of elementary particles and fields; nuclear, atomic, and molecular physics; classical areas of phenomenology and their applications; fluids, plasmas, and electrical discharges; condensed matter; cross disciplinary physics and related areas of science and technology; and geophysics, astronomy, and astrophysics are among the subjects listed in the Table of Contents.

Scope: It is stated in the preliminary pages that the index is "...photocomposed from a computer file of bibliographic information about the articles published by the Institute and its member societies." It is also noted that each issue "...contains approximately 4,000 abstracts of articles published in the primary journals in one quarter." International material is covered.

Locating Material: Entries are numbered sequentially. These numbers are referred to in the indexes.

Abstracts: Titles and abstracts are in English and the senior author's affiliation is given. Article titles are repeated under up to three other subjects with a reference to the abstract number.

Indexes: An author index is found in each issue which cumulates annually. There is also an annual subject index.

Other Material: Extensive explanatory material, which includes an outline of the classification scheme used, is provided.

Periodicals Scanned: A list is in each issue and in the annual indexes.

Database: Entries are available through SPIN from 1975 and are updated monthly. Abstracts and keywords are provided.

ELECTROANALYTICAL ABSTRACTS, 1963-
Birkhäuser Verlag, Bimonthly. Bears subtitle: International Journal Dealing with the Documentation of All Aspects of Fundamental Physico-Chemical and Analytical Electrochemistry. Continues the abstract section of the Journal of Electroanalytical Chemistry.

Arrangement: Arranged by broad subjects.

Coverage: Documents all aspects of fundamental physicochemical and analytical electrochemistry. The abstracted papers are divided into the following classes: Fundamental electrochemistry, Apparatus, Voltammetry, Palarography, Amperometry, Polentiometry, Conductometry, Electroanalysis, Coulometry, Electrophoresis, Other Methods, Related Topics, Book Reviews and New Books.

Scope: International.

Locating Material: Entries are in numerical order through the volume. Book reviews and new books are in alphabetical order within each issue.

Abstracts: Abstracts are in English (formerly in English, French, and German). The language of the article is specified. Abstractor's name, address of the author, and number of references are indicated. At the end of each subject area is a list of abstract numbers for related subjects.

Indexes: Author and subject indexes for each volume (which appear in issue number six) are provided.

Other Material: A list of symbols used in the abstracts is found on the inside back cover.

JOURNAL OF CURRENT LASER ABSTRACTS, 1967-
Institute for Laser Documentation. Monthly. Formerly: Laser/Maser International, 1964-66.

Arrangement: Arranged by broad subjects subdivided into narrower topics.

Coverage: "...laser theory, design and application ... includes periodicals, conference proceedings, government reports, patents, dissertations, books, etc., in a research and technology range extending from the theory of electromagnetic radiation to such diverse applications of lasers in guidance, holography, display, geodesy, medicine, ophthalmology, laser fusion, and laser isotope separation" (from a statement in the preliminary pages).

Scope: International.

Locating Material: Within each subject subdivision, entries are alphabetical by title.

Abstracts: Titles and abstracts are in English. Language of the articles is noted and the location of the author is given.

Indexes: An author (both personal and corporate) index is in each issue which cumulates annually. An annual subject classification index is provided.

Other Material: A list of new laser patents is in each issue.

NUCLEIC ACIDS ABSTRACTS see BIOCHEMISTRY ABSTRACTS

PHYSICS ABSTRACTS see SCIENCE ABSTRACTS

PHYSIKALISCHE BERICHTE, 1920-1978.
Deutsche Physikalische Gesellschaft, Verlag Friedr. Vieweg und Sohn, GmbH. Ceased. Suspended from 1944-1947, but Vol. 28 (1949) includes papers of the war and post-war years. Formed by the union of Fortschritte der Physik, 1845-1918, Halbmonatliches Literaturverzeichnis, 1902-1919, and its Beiblätter to the Annalen der Physik, 1877-1919.

Arrangement: Arranged by broad subjects subdivided into narrower topics.

Coverage: All aspects of physics including relevant material in geophysics, astrophysics and biophysics.

Scope: International.

Locating Material: Entries are numbered consecutively within each issue. Each abstract carries a unique number made up of the issue number and the consecutive number: e.g., 2-99 refers to the 99th abstract in issue number 2 of a given volume.

Abstracts: Titles are in the original language and abstracts are in German. Location of the senior author is given and the abstractor's name is provided. Price and format information is supplied for monographs.

Indexes: An author index is in each issue which cumulates annually and a cumulated classified index for each volume is furnished.

Other Material: A list of abstractors, an outline of the classification scheme, a list of key words and a list of abbreviations appear in the annual index volume and in the January and July issues.

Periodicals Scanned: A list appears in the annual index volume and semiannually in the January and July issues. The full title, abbreviations, and address are given.

RHEOLOGY ABSTRACTS, A SURVEY OF WORLD LITERATURE, 1958-
Pergamon Press, Ltd. Quarterly. Supersedes the abstracting section of the British Society of Rheology Bulletin.

Arrangement: Arranged by broad subjects subdivided into narrower topics.

Coverage: All aspects of the subject including theories, instruments and techniques, solids, polymers and other viscoelastic materials, solutions, pastes and suspensions, liquids, and general information are covered.

Scope: International.

Locating Material: Entries are numbered consecutively through the volume.

Abstracts: Titles and abstracts are in English. The language of the article is indicated. The address of the first author and the abstractor's initials are given. If the author's abstract is used or is abridged, this is noted. Some abstracts are followed by a series of numbers; these refer to systems of classification of papers on non-Newtonian Flow and Solids.

Indexes: An annual author and subject index is provided.

Other Material: Book reviews and summaries of papers to be read at forthcoming meetings appear occasionally. A discussion of the non-Newtonian Fluids Classification System and the Solids Classification System appears at intervals.

Periodicals Scanned: A list of journals covered, giving full title and abbreviation, is sometimes included in the first issue of, the volume.

SCIENCE RESEARCH ABSTRACTS JOURNAL, 1972-
Cambridge Scientific Abstracts. Bimonthly. Issued in two parts: A) Superconductivity; Magnetohydrodynamics and Plasmas; Theoretical Physics and, B) Laser and Electro-Optic Reviews; Quantum Electronics; Unconventional Energy Sources.

Arrangement: Arranged by subject. Broad areas are subdivided into smaller categories.

Coverage: All aspects of the subjects listed as headings for the two parts are thoroughly indexed. The Table of Contents lists the subdivisions under the broader headings.

Scope: International.

Locating Material: The abstracts are numbered consecutively through each volume. The number has seven digits. The first two indicate the year and the last five are the sequential number of the abstracts; these begin with 00001 in each volume.

Abstracts: Titles are in English and usually in the language of the article if it is other than English. That language and summaries in other languages are indicated. Up to ten authors' names are given and the senior author's affiliation is furnished.

Indexes: Author and subject indexes are found in each issue. These cumulate annually.

Other Material: Brief explanatory material is found in the preliminary pages which includes a list of abbreviations for languages.

Periodicals Scanned: A list is in the first issue of each volume.

SOLID STATE ABSTRACTS JOURNAL, AN ABSTRACT JOURNAL INVOLVING THE PHYSICS, METALLURGY, CRYSTALLOGRAPHY, CHEMISTRY AND DEVICE TECHNOLOGY OF SOLIDS, 1960-
Cambridge Scientific Abstracts, Inc. Ten issues per year. Supersedes: Semiconductor Electronics 1957-1959/60. Formerly: Solid State Abstracts...., 1960-1965.

Arrangement: Arranged by broad subject headings with subheadings, the latter being designated by a special classification scheme. For example: under the broad subject heading, "Solid State Physics," there appear the letters SSP1 (Atoms and Nuclei in Solids), under this heading is SSP1.1 (Ionic Spectra), which in turn is divided into SSP1.1.2 (General), SSP1.1.3 (Crystal Fields), etc.

Coverage: Complete and comprehensive coverage of the theory, production, and application of solid state materials and devices. Covers in addition to periodical articles, government reports, conference proceedings, books, dissertations, and patents.

Scope: International.

Locating Material: Entries are numbered consecutively through each volume. The number has seven digits and the first two indicate the year.

Abstracts: Full bibliographical citation is provided; up to ten authors are given and the address of the first author is supplied. Titles and abstracts are in English with an indication of the language of the article if it is other than English. The titles may also appear in the original language. Abstracts vary in length and occasionally none is provided.

Indexes: Monthly issues have author and subject indexes which cumulate annually.

52 / Abstracts and Indexes

Other Material: The Table of Contents is an outline of the subjects covered, arranged by classification number. Some issues have an abridged Table of Contents for Solid State Abstracts on Cards and some have a sample page from the Classification System for the "Physics" Section and the "Circuits" Section of Solid State Abstracts on Cards.

Periodicals Scanned: The list appears in the Source Index in issue number one of each volume. It gives full titles, abbreviations (as used in the abstracts), and the name and full address of the publisher.

SPECTROCHEMICAL ABSTRACTS, 1930/37-1973.
Adam Hilger, Ltd. Annual since 1962/63; prior to that time, one volume covered several years. Ceased.

Arrangement: Arranged by broad subjects subdivided into narrower topics.

Coverage: All aspects of spectrochemistry including substance analyzed, apparatus, methods and basic theory are covered. Included is a section on books and reviews.

Scope: International.

Locating Material: Entries are numbered consecutively from the first volume issued.

Abstracts: Titles and abstracts are in English. Abstracts are usually brief but informative and the bibliographic citation is complete.

Indexes: An author index and an Index of Elements are found in each issue.

Periodicals Scanned: To be included in Volume 18.

THEORETICAL CHEMICAL ENGINEERING ABSTRACTS, 1964-
Theoretical Chemical Engineering Abstracts, Six issues a year; one issue covers two months.

Arrangement: Arranged by broad subjects subdivided into narrower topics. "See also" references at the end of sections lead the reader to alternate terms and entry numbers.

Coverage: All aspects of theoretical chemical engineering including fluid dynamics, head transfer, mass transfer, kinetics and thermodynamics, chemical reactor engineering, design, control, physical separation, crushing, pumps, compressors, monograms and general information are included. Also covers books, government reports, conference proceedings, etc.

Scope: International.

Locating Material: Entries are numbered consecutively through the volume.

Abstracts: Titles and abstracts are in English. The language of the article is indicated and the number of references is given. If available, price is given for monographic works.

Indexes: An annual subject index which lists, under the subject headings, abbreviated titles of all articles cited that may be classified under that particular subject.

Periodicals Scanned: A list is provided with the annual index.

NUCLEAR SCIENCE AND SPACE SCIENCE

INIS ATOMINDEX: AN INTERNATIONAL ABSTRACTING SERVICE, 1970-
International Atomic Energy Agency: distributed by Unipub. Semimonthly. Nuclear Science Abstracts was discontinued because this publication contained the same material. Supersedes: List of References on Neuclear Energy, 1959-68.

Arrangement: Arranged by broad subjects subdivided into narrower topics. Each broad subject is indicated by a letter and two zeros (A00 - Physical Science) and the subdivisions are given that letter and a two-digit number beginning with "10." This is shown in the Table of Contents. Within each subject the references are arranged by type of literature: first is technical reports, patents, and conference papers, followed by journal articles and books.

Coverage: Publications related to nuclear science and its peaceful applications. Among the subjects covered are physics, chemistry, earth sciences and life sciences, isotope and radiation applications, nuclear reactors, waste management, safeguards and inspections, and mathematical methods and computer codes.

Scope: International.

Locating Material: Entries are numbered consecutively from Volume one. These numbers (referred to as reference numbers or "RN") are used as access from the indexes.

Abstracts: Titles and abstracts are in English. The titles are repeated in the language of the article if it is other than English and the language is indicated. If no abstract is provided, a set of descriptors will be given. Additional subject headings are noted. All authors' names are given and the affiliation of the senior author is specified. Corporate sources are also noted. Report numbers and availability are given if this is important.

Indexes: Personal author, corporate source, subject, conference (by date and by place) and a report and patent number index are provided in each issue. All of these cumulate semiannually. All authors are included. "See" and "see also" references lead the reader to alternate terms in the subject index; titles and reference numbers are given except in the report number index where only the reference number is provided.

Other Material: Very detailed explanations of how to use the journal and its indexes are found in the preliminary pages of each issue. Of particular interest is a description of how to trace conference papers. A list with addresses of INIS (International Nuclear Information System) liaison officers is provided. Availability and order information and a price code are provided.

Database: INIS has been online from 1970; abstracts have been provided from 1976. Keywords are available and the update is twice per month.

INTERNATIONAL AEROSPACE ABSTRACTS, 1961-
Technical Information Service, American Institute of Aeronautics, Inc. Semimonthly except June which has three issues.

Arrangement: Arranged by broad subjects subdivided into narrower categories.

Coverage: Monographs, theses, conference proceedings, meeting papers and journal articles in the field of aeronautics and space sciences are found. Broad subjects include aeronautics, astronautics, chemistry and materials, geosciences, mathematical and computer sciences, physics, and space sciences.

Scope: International.

Locating Material: Each entry is assigned an identification number which consists of the prefix A, the year, a dash and a five-digit number (these are not sequential). An asterisk attached to the number indicates NASA sponsorship; a # indicates availability on microfiche. These numbers are referred to in all indexes.

Abstracts: Abstracts are in English. Titles are in English and in the language of the article. When not in English the original language is noted. Authors and affiliations, number of references, and contract numbers are provided.

Indexes: Subject, personal author, contract number, meeting paper, report number and accession number indexes are in each vidual issue indexes give annually. Index entries in the individual issue indexes give category location; in the cumulated index, page location is given. Explanatory notes are provided with each index to facilitate its use.

Other Material: There is a general explanation in the preliminary pages of each issue. Illustration of sample citation and instructions on availability and price of material are also supplied. A number of bibliographies are published by NASA as Special Publications (SuDocs number: NAS 1.21:SP #) which provide annotated references to reports and articles announced in Scientific and Technical Aerospace Reports or in International Aerospace Abstracts. Abstracts and citations appear as they did in STAR or IAA and carry the original accession number. These include Aerospace Medicine and Biology, Aeronautical Engineering, NASA Patent Abstracts, Earth Resources, and Energy, a Continuing Bibliography.

Database: Entries are available through NASA from 1961 with weekly updates.

NUCLEAR SCIENCE ABSTRACTS, 1948-1976.
Atomic Energy Commission. Preceded by: Abstracts of Declassified Documents. The Energy Research and Development Administration took over the indexing of this material in ERDA Research Abstracts which later became Energy Research Abstracts.

Arrangement: Arranged by subject. There are subtopics under broad headings.

Coverage: Literature of nuclear science and technology including chemistry, engineering, earth sciences, nuclear materials and waste management, physics, reactor technology, etc. is covered (see Table of Contents). Patents, conference papers and similar material are included.

Scope: International.

Locating Material: Entries are numbered consecutively through each volume. Index entries refer to volume number and abstract number.

Abstracts: The titles and abstracts are in English. The language of the article is indicated. All authors' names are recorded and the senior author's affiliation is given. The source of the abstract is noted. "See references" guide the user to alternative or additional citations. Patent numbers, report numbers, and other articles of information to facilitate acquiring a document are provided.

Indexes: Corporate author, personal author, subject and report number indexes are in each issue. The report number index includes patents and conference papers. The author and subject indexes for each volume and the report number index cumulate on a calendar year basis. Explanatory material at the beginning of each index facilitates its use. Cumulative indexes are available as follows: Vol. 1-4, and 5-10, subject and authors; Vol. 11-15 and 16-20, subject and personal and corporate authors; Vol. 1-15, 16-20 and 21-26, report number index. An index for Vol. 21-25 (subject, personal and corporate authors) is available separately (from University Microfilms).

Other Material: A list of report collections in the United States by state is found on the verso of the front cover and a list of collections in other countries by country is on the verso of the back cover. There is extensive explanatory material in the preliminary pages and in the appendices. Here is found information on the subject matter and arrangement. There is a separate index to this material.

Periodicals Scanned: A complete list of periodicals abstracted and indexed appears in the issue of January 15. Supplements appear in issues dated April 15, July 15, and October 15. This gives abbreviated and full title, and country of origin. If foreign language journals are available in English, this is noted. Explanatory material at the beginning of the list facilitates its use.

SCIENTIFIC AND TECHNICAL AEROSPACE REPORTS, 1963- National Aeronautic and Space Administration, (U.S. Government Printing Office). Semimonthly. Supersedes: Technical Publications Announcements.

Arrangement: Arranged by subject categories.

Coverage: Reports, dissertations, NASA patents and patent applications in aeronautics, space and supporting disciplines including aerodynamics, geophysics, meteorology, propulsion systems, space sciences, etc. are found (see Table of Contents).

Scope: International.

Locating Material: Each entry is assigned an identification (accession) number consisting of the letter N, two digits for year, a dash and a five-digit number beginning with 10,000. The five-digit numbers currently run consecutively through each volume. Thus, N83-10002 is the second entry of 1983. The accession number is referred to in all indexes. The number will sometimes have symbols following it; e.g.: *, indicates a NASA-sponsored document; #, indicates availability on microfiche; +, indicates microfiche unavailable but a one-to-one facsimile can be provided. An availability statement is given with each entry with an explanation provided in the preliminary pages of each issue.

Abstracts: Titles and abstracts are in English. Titles are also in the language of the report. The language is noted as well as summaries in translation. References are indicated (but not how many) and the source of the abstract is given. All authors' names, the corporate source, contract or grant number, report numbers, the availability source, and the COSATI (Committee on Scientific and Technical Information) code are indicated. If a citation is a U.S. patent, number and patent class numbers are given. "See references" lead the searcher to related subsidiary citations.

Indexes: Each issue has a subject, personal author, corporate source, contract number and report/accession number index. These cumulate semiannually and annually. Explanatory notes at the beginning of the indexes facilitate their use.

Other Material: The introduction provides a source of information about STAR as well as an explanation about the availability of the material. A list of organizations from which reports are available and their addresses, and a detailed explanation of the entries are given. A brief list of institutions with collections of NASA documents is provided on the inside back cover of each issue and in the cumulative indexes. A number of bibliographies are published by NASA as Special Publications (SuDocs number: NAS 1.21:SP #) which provide annotated references to reports and articles announced in Scientific and Technical Aerospace Reports or in International Aerospace Abstracts. Abstracts and citations appear as they did in STAR or IAA and carry the original accession number. These include Aerospace Medicine and Biology, Aeronautical Engineering, NASA Patent Abstracts, Earth Resources, and Energy, a Continuing Bibliography.

Database: Online bibliographic services are available through NASA from 1962. There are weekly updates with abstracts and keywords provided. International Aerospace Abstracts is also in this database.

EARTH SCIENCES, ARCHAEOLOGY, AND ANTHROPOLOGY

ABSTRACTS IN ANTHROPOLOGY, 1970-
    Baywood Publishing Company. Quarterly.

Arrangement: Arranged by broad subject areas in anthropology. Within each area, abstracts are arranged alphabetically by senior author with cross references from joint authors.

Coverage: Archaeology, ethnology, linguistics, physical and cultural anthropology are covered.

Scope: International.

Locating Material: In Volume 1, the abstracts are numbered consecutively in each issue. From Volume 2, they are consecutive through the volume.

Abstracts: Abstracts range in length from very brief to quite long and are in English. The titles are in the language of the article and that language is named.

Indexes: An author and subject index is provided in each issue.

Other Material: Occasional notes from the editor indicate changes or projected changes in the publication. "Instructions to Contributors" appears from time to time.

Periodicals Scanned: A list is found in each issue.

ABSTRACTS OF NORTH AMERICAN GEOLOGY, 1966-1971.
    U.S. Geological Survey. Ceased. The intention was for this publication to amplify the Bibliography of North American Geology. Superseded: Geological Abstracts, 1953-1958 and Geo-Science Abstracts, 1959-1966.

Arrangement: Alphabetical by author (or title of journal if no author is given).

Coverage: Technical material relating to North American geology including geomorphology, exploration, minerals, paleogeography, oil and gas fields, wells and drill holes, etc., is covered (see Subject Index).

Scope: International in that articles of a general nature by North American authors published in foreign journals are included. Foreign authors are covered only if the material appears in a North American journal.

Locating Material: Each entry has an identification number which is referred to in the subject index. One can also seek the author's name in the alphabetical list.

Abstracts: The titles and abstracts are in English. All authors are given, but "see references" refer from joint authors to senior author for complete citation and abstract. Illustrative material is noted. The abstractor's initials are given or if the author's abstract is used, this is noted. The abstracts are in general quite informative.

Indexes: A subject index is found in each issue.

Periodicals Scanned: A list of serial publications cited for the first time is provided in each issue. A list of journals commonly cited is available separately.

ALLOYS INDEX see METALS ABSTRACTS

ALPHABETIC SUBJECT INDEX TO PETROLEUM ABSTRACTS see PETROLEUM ABSTRACTS

BIBLIOGRAPHIE DES SCIENCES DE LA TERRE see BULLETIN SIGNALETIQUE

BIBLIOGRAPHIE DES SCIENCES GEOLOGIQUES see BULLETIN SIGNALETIQUE

BIBLIOGRAPHIE GEOGRAPHIQUE INTERNATIONALE, 1891- Paris, Centre National de la Recherche Scientifique. Annual. V. 1-24 issued with Annales de Géographie. Title and publisher vary.

Arrangement: Divided into two parts: A) Topical Section which is by subject; and B) Regional Section which is by region of the world (Europe, Asia, etc.), then by country within a region and in some cases by section of the country. Within each subpart, the arrangement is alphabetical by author.

Coverage: The subjects covered include teaching, history of geography, cartography, geomorphology, hydrology, biogeography, population geography, economic and industrial geography, political geography (third world) and physical geography. Monographs, conference proceedings, etc., are included with the journal articles.

Scope: International.

Locating Material: The Table of Contents aids the user in finding the correct subject section and is located in the back of the volume. Each section is identified by a capital letter, a Roman numeral and a small letter. The entries within each section are numbered consecutively. These letters and numbers are referred to in the index.

Abstracts: Generally the titles are in the language of the article but for some languages (e.g., Russian) they are repeated in French. The abstracts are in French and vary in length from one line to half a page or more. Occasionally none appears. Frequent "see references" and notations of related articles lead the reader to alternate sources.

Indexes: An author index is in each volume. Some sections have their own subject index but this is not provided for all of them.

Other Material: A list of contributors and a list of abbreviations used are provided in each volume.

BIBLIOGRAPHY AND INDEX OF GEOLOGY, 1969-
American Geological Institute. Monthly with annual compilations. Succeeded the Bibliography of North American Geology (see separate entry) and Bibliography and Index of Geology Exclusive of North America which covers the literature from 1933 through 1968.

Arrangement: The monthly issues have four sections: Serials, Fields of Interest, Author, and Subject Index. The Fields of Interest section contains the bibliographic citations of source material. These are arranged by subject (29 broad areas). In the annual volumes the bibliography section is alphabetical by author and the index section is alphabetical by subject (using terms from the monthly issues).

Coverage: The "earth science literature of the world" is indexed, including area, economic, engineering, and extraterrestrial geology, geochemistry, geomorphology, marine geology, paleobotany, sedimentary petrology, hydrogeology and hydrology, various aspects of surface geology and structural geology, etc. Books, serial, reports, maps, and North American theses are included. (See Table of Contents of the monthly issues.)

Scope: International.

Locating Material: Each entry in the monthly issues is numbered consecutively through the volume. The indexes refer to these numbers. In the annual compilation the bibliography is arranged alphabetically by author and the index, alphabetically by subject using key words. These may be primary, secondary, or tertiary terms. Further, a word may be a primary term and appear again in a secondary position under another primary heading. These entries do not provide complete bibliographic citation as the bibliography section does. However, the author is given and reference is made to issue by month, subject section, and accession number.

Abstracts: Not usually applicable, but provided for some citations. Titles are given in English and in the language of the article. The language is sometimes noted as are summaries in other languages. Illustrative material and the number of references is indicated. All authors' names are given and "see references" refer from joint author to primary author. Senior author's affiliation is given when available.

Indexes: Each monthly issue contains a subject and an author index. The index to the annual compilation refers to the Bibliography Section by author.

Other Material: Explanatory information appears in the Preface of the monthly issues and in both sections of the annual compilation.

Periodicals Scanned: A list is in each monthly issue and in the bibliographic section of the annual compilation. Abbreviation, full title and place of publication are given. A separate section lists "Special Publications" (festschrifts are included here). GeoRef Serials List and KWOC Index, a list which includes all serials indexed since 1967 and a KWOC index to significant words in the titles are available separately.

Database: Citations are available through GEOREF from 1961 and are updated monthly. Keywords are provided and occasionally there will be abstracts.

BIBLIOGRAPHY AND INDEX OF MICROPALEONTOLOGY, 1972-
Micropaleontology Press of The American Museum of Natural History in cooperation with the American Geological Institute. Monthly.

Arrangement: Arranged by microfossil groups. Within each group the arrangement is alphabetical by author's name.

Coverage: The citations are in twelve categories which include
Algae, Annelida, Calcareous Nannofossils, Conodonts, Diatoms,
Foraminifera, Ostracoda, Polunomorphs, Protista, and Radi-
olaria.

Scope: International.

Locating Material: The categories are numbered 01 through 12 and
then the articles are arranged alphabetically by the author's
name. The index refers to the category number. To find a
citation by a particular author, turn to the proper category
and search alphabetically.

Abstracts: Not applicable. Titles are in English and in the language of the article if other than English. Only the first three
authors are listed in the citations but cross-references to
all authors are provided. Availability information is given
if this is not immediately apparent in the bibliographic data.

Indexes: There is an author index in each issue which cumulates
annually.

Other Material: A brief explanation of the arrangement and format
of the Index is provided in the preliminary pages.

BIBLIOGRAPHY OF NORTH AMERICAN GEOLOGY, 1785/1919-1970.
U.S. Geological Survey. Ceased. Published as U.S. Geological Survey Bulletin (which continues). List of Bulletin
numbers in introduction of each volume. The original volume,
Bulletin 746, covers the literature from 1785-1918 and was
published in 1923, compiled by John M. Nickles and titled
Geologic Literature of North America.

Arrangement: Arranged alphabetically by author. Each entry is
assigned an accession number. They are not arranged in
the volume sequentially.

Coverage: The geology of the North American continent, Greenland,
West Indies, Hawaii, Guam and other island possessions is
covered. The range of subject matter is extremely wide.
The user is referred to the subject index in each volume.

Scope: Foreign journals are included when articles are by American
authors and deal with North American localities or are of a
general nature, but not included if they deal only with foreign
areas. Articles by foreign authors are included if they appear in North American journals or if they are on North
America (regardless of the place of publication).

Locating Material: Although numbers are assigned to the abstracts
they are only for reference in using the index, they are not

sequential. The searcher either takes an author approach and searches through the alphabetical listing or uses the subject index.

Abstracts: Not applicable. Entries are in English and the language of the article. The language is noted as are summaries in other languages. Illustrative material is noted. All authors' names are given and cross references are made from joint authors to primary author.

Indexes: A subject index is found in each volume which gives brief title, author and accession number.

Other Material: Brief explanatory information is in the introduction. A list of U.S. Geological Survey Bulletin numbers (and the year) which are the Bibliography of North American Geology is also provided.

Periodicals Scanned: A list is found in each volume which gives full title, abbreviations and place of publication.

Database: Entries from this bibliography are available through GEOREF.

BIBLIOGRAPHY OF SEISMOLOGY, 1965-
International Seismological Center. Semiannual.

Arrangement: Divided into three sections: 1) subject index, 2) author index, and 3) list of citations to important events. In the subject sections, the arrangement is alphabetical by author under each heading. The author section is alphabetical by author and the event section is arranged by the name of the event from the most recent backward in time for five years.

Coverage: Seismological literature which appeared in the six-month period covered by each issue, and earlier material that may have become available is included. Among the subjects are such terms as absorption, acceleration, acoustics, aftershock, borehole, seismic sounding, etc. Geographic locations and place names are included as topics.

Scope: International.

Locating Material: The researcher may use an author or subject approach. Also the name of an "important event" could be used as an access to papers.

Abstracts: Not applicable. Complete bibliographic information is given to papers in all three sections. The titles are in English and if the article is in another language, that language is indicated in parentheses. Abstracts in other languages are also noted.

Other Material: A list of abbreviations for journals is found on the inside front and back cover of each issue. Brief explanatory information is found in the Introduction.

Periodicals Scanned: See "Other Material" above.

BRITISH ARCHAEOLOGICAL ABSTRACTS, 1968-
   Council for British Archaeology. Semiannual.

Arrangement: Arranged by subject within the period. There is a "section finder" to aid the researcher.

Coverage: Archaeology in Great Britain and Ireland. The Preface states that "The purpose of British Archaeological Abstracts is to indicate the most significant material currently being published." It also provides the research source for consultation in years to come, and enables specialists to discover what is happening in fields other than their own. Monographs, corpora, gazetteers, and book reviews are provided.

Scope: International.

Locating Material: The abstracts are numbered consecutively. The number begins with two digits which are the last two numbers of the year and then the sequential number. The "section finder" leads the searcher to material by means of a grid with code numbers and letters. Abstracts pertaining to a particular subject and period are given the appropriate code or codes (which are found in the top line of its headings).

Abstracts: Titles are in the language of the articles and the abstracts are in English. Illustrative material is indicated; if there are ten or more references, the notation "refs." is found.

Indexes: Annual subject index is provided.

Other Material: A detailed description of this publication is given which includes a list of contributors, a description of entries, and a list of abbreviations.

Periodicals Scanned: A list is provided in each issue.

BRITISH GEOLOGICAL LITERATURE, 1964-68; 1972-   (new series).
   Bibliographic Press, Ltd. Quarterly.

Arrangement: Arranged by subject.

Coverage: Mineralogy; history; marine and regional geology; geophysics; geochemistry; crust/continents; external processes, geomorphology; climate, stratigraphy, paleogeography; petrology; economic geology; palaeontology, mining production; geotechnics are listed as major topics. Coverage is limited to the "British Isles and adjacent sea areas" (cover). Books, proceedings, and reports are included.

Scope: International.

Locating Material: Entries are numbered consecutively through each volume. The first two digits indicate the year. The index reference is to these numbers.

Abstracts: Brief annotations are provided. Titles and annotations are in English. The number of references and illustrative material are given and the senior author's affiliation is noted.

Indexes: An author index is found in each issue. Annual subject, locality, and author indexes, and a regional guide are provided.

Other Material: A list of cross references is supplied.

Periodicals Scanned: A list is provided from time to time. The last found at this writing was in 1980, Part 4.

CADMIUM ABSTRACTS, 1977-
    Cadmium Association (London)/ Cadmium Council (New York). Quarterly.

Arrangement: Arranged by subject.

Coverage: This is a survey of current world literature on the properties and uses of cadmium and its alloys and compounds. The subjects include analysis, batteries, biochemistry, chemistry, coatings, corrosion, economics and statistics, electrochemistry, electronics, environment, extraction and refining, health and safety, physical metallurgy, plating, pollution control, recycling, rubber and plastics, solar cells and tribology.

Scope: International.

Locating Material: Each entry is given an alphanumeric designation. The first letter indicates the bulletin (C, Cadmium Abstracts; Z, Zinc Abstracts; L, Lead Abstracts), the next two figures stand for the year of publication and then the abstract number which is used in the index. Also on the top line of each abstract is a classification code (CC) number (abstracts are listed in this order), a code for document type (DT), and a language code (LC).

Abstracts: Titles are in English and in the language of the article. Illustrative material is indicated and the language of the article is noted. Patent numbers are supplied where applicable. Frequent "see references" lead the searcher to related items.

Indexes: Annual author, subject, standards, conference and patent indexes are provided.

Other Material: A list for the classification code, code for document type and language code is provided with the first issue of each year. Additional copies are available on request. An informative overview of some of the material covered is found in each issue.

Database: Entries are available through ZLC ABSTRACTS.

GEO ABSTRACTS, 1960-
Geo Abstracts, Ltd. Published in seven sections each appearing six times a year. The Sections are: A - Landforms and the Quarternary; B - Climatology and Hydrology (formerly Biogeography and Climatology); C - Economic Geography; D - Social and Historical Geography (formerly Social Geography and Cartography); E - Sedimentology; F - Regional and Community Planning; G - Remote Sensing, Photogrammetry and Cartography. Formerly: Geomorphological Abstracts, 1960-1965; GEOGRAPHICAL ABSTRACTS, 1966-1971. The description below pertains to all sections.

Arrangement: Arranged by subject.

Coverage: The subject matter indicated by the titles of the various sections receives a fairly wide coverage. Among the subjects are: ecology, hydrology, soil mechanics, regional physiography, conservation, population distribution, instrumentation and techniques, economic geography, man and enviroment, paleogeography and tectonics, urban and rural planning, statistical cartography, meteorology and forecasting techniques, and landforms. (See Contents of each section.) Books, proceedings, reports, etc. are included.

Scope: International.

Locating Material: Each entry has an identification consisting of two numerals indicating the volume year, a letter indicating the subject section, a slash and a four digit number which begins 0001 and runs sequentially through the volume. Entries in the indexes refer to these numbers.

Abstracts: The abstracts are in English. The titles, wherever possible, are in the language of the article and in English and sometimes only in English. The language of the article and summaries in other languages are noted. Illustrative

70 / Abstracts and Indexes

material and the number of references are indicated and all authors' names are given. The abstractor's name or the source of the abstract is indicated.

Indexes: Annual author and regional indexes are found in each section. From 1972 the annual indexes are published in two parts; Part I, covering sections A, B, E, and G; Part II covering sections C, D, and F. Cumulative subject and author indexes are available as follows: Geomorphological Abstracts, 1960-1965; Geographical Abstracts Sections A-D, 1966-1970, 4 volumes; Geo Abstracts Sections A-D, 1971-1975, 4 volumes; Geo Abstracts Sections E-F, 1972-1976, 2 volumes; Geo Abstracts Section G, 1974-1978.

Other Material: Occasionally there will be an editorial, usually in the first issue of a section. A description of how to use the annual indexes appears in each volume.

Periodicals Scanned: The annual index carries a list of journals scanned in that year. The two volume annual indexes from 1972 carry the list of journals for the relevant parts.

Database: A note in the 1979 annual index (published in 1981) states that the indexes are held on magnetic tape and from 1981 Geo Abstracts will be available on magnetic tape through one or more online systems (including Lockheed Dialog).

GEOPHYSICAL ABSTRACTS, 1929-1971.
U.S. Geological Survey. Ceased.

Arrangement: Arranged by subject.

Coverage: Age determinations, earthquakes, geodesy, gravity, isotope geology, magnetic surveys, scientific exploration, submarine geology, volcanology, etc. are covered (see Table of Contents).

Scope: International. A note in the Introduction explains that ordinarily material with limited circulation is not included, nor are papers presented orally at meetings.

Locating Material: Each entry has an abstract number consisting of the volume number and sequential numbers which begin in each issue with 001. These numbers are referred to in the subject index. Only the sequential numbers are given in the author index.

Abstracts: Titles are in the language of the article and in English. Abstracts are in English. All authors are given and illustrative material is noted. Abstracts are sometimes quite concise but generally informative. Abstractor's initials are given. (A list of full names is provided.) If the initials follow the notation "author's abstract," this is an indication of a translation.

Indexes: Author and subject indexes are found in each issue. These cumulate annually.

Other Material: Explanatory material is found in the "Introduction."

Periodicals Scanned: A list of journals commonly cited is available separately from the U.S. Geological Survey. Each issue since no. 224 gives supplements to the master list.

Database: Entries from this source are available through GEOREF.

GEOPHYSICS AND TECTONICS ABSTRACTS, 1977-
   Geo Abstracts, Ltd. Six times a year. Formerly: Geo-Physics Abstracts 1977-1982 (not to be confused with Geophysics Abstracts, 1958-1971 published by the U.S. Geological Survey).

Arrangement: Arranged by broad topics subdivided into narrower subjects.

Coverage: Age determination, cosmogeny, electrical properties, gravity, heat and heat flow, magnetics, rheology, radioactivity, seismology, structural geology, tectonics, regional geology, and volcanology are the broad topics covered.

Scope: International.

Locating Material: Abstracts are numbered consecutively through each volume beginning with 0001 in issue number one. The numbers consist of two digits indicating the year, the letter "p," a slash (/) and then the consecutive number.

Abstracts: The titles are in English and in the language of the article. The abstracts are in English. The number of references and illustrative material is noted. The language of the article is usually indicated as are English summaries if they are provided.

Indexes: Annual author, subject, and regional indexes are found.

Other Material: Occasionally there will be editorial comments or notices of publications. An explanation of the annual index is given.

Periodicals Scanned: A list is published with the annual index.

IMM ABSTRACTS, 1950-
   Institute of Mining and Metallurgy. Bimonthly.

Arrangement: Arranged by subject (the Universal Decimal Classification scheme is employed). Most divisions have subtopics.

Coverage: The broad subjects listed include minerals industry, mathematical methods and computing, economic geology, mining, mineral processing, metallurgy, environment, general and civil engineering.

Scope: International. A note in the introductory pages states that IMM Abstracts is prepared "...from published material received by the Institution's Library."

Locating Material: A seven-digit number is used which incorporates two digits indicating the year and then a consecutive number.

Abstracts: Not all entries include abstracts. Those that are provided are in English as are the titles. The language of the article is indicated and availability of the original material is provided. The Universal Decimal Classification number is given.

INTERNATIONAL PETROLEUM ABSTRACTS, 1969-
   Heyden and Son, Inc. Quarterly. Formerly: Institute of Petroleum Abstracts, 1969-1972.

Arrangement: Arranged by broad subjects subdivided into narrower topics.

Coverage: Oilfield exploration and exploitation, transport and storage, refinery operations, products, corrosion, engines and automotive equipment, safety, economics and marketing, pollution and oil firms are covered. (See Table of Contents for subdivisions.)

Scope: International.

Locating Material: Entries are numbered consecutively through the volume.

Abstracts: Titles and abstracts are in English. The language of the article if it is other than English is indicated. The abstracts are usually brief, but may be fairly long. If they are based on the author's abstract, this is indicated; otherwise, the abstractor's initials are given.

Indexes: An author index is found in each issue. These cumulate annually and there is an annual subject index.

Earth Sciences, Archaeology, and Anthropology / 73

Other Material: A list of abstractors appears annually.

Periodicals Scanned: The list, which gives full title and abbreviations, is provided annually.

LEAD ABSTRACTS, 1958-
  Lead Development Association. Quarterly.

Arrangement: Arranged by six main subject headings with subsections.

Coverage: This is a survey of current world literature on the properties and uses of lead and its alloys and compounds. Economics and statistics, metal extraction, uses and processes, pollution control, environmental aspects, health and hygiene, biochemistry, properties of materials and their management are among the topics covered.

Scope: International.

Locating Material: Each entry is given an alphanumeric designation. The first letter indicates the bulletin (C, Cadmium Abstracts; L, Lead Abstracts; Z, Zinc Abstracts and the next two figures stand for the year of publication and then the abstract number which is used in the index is given. Also on the top line of each abstract is a classification code (CC) number (abstracts are listed in this order), a code for document type (DT) and a language code (LC).

Abstracts: The titles are in English and in the language of the article. Illustrative material is indicated and the language of the article is noted. Patent numbers are supplied where applicable. Frequent "see references" lead the searcher to related items.

Indexes: Annual author, subject, patent, standards, and conference indexes are provided.

Other Material: A list of the Classified Code, code for Document Type, and Language Code is provided with the first issue of each year. Additional copies are available on request. An introduction gives an informative overview of some of the material covered in each issue.

Database: Entries are available through ZLC ABSTRACTS.

METALS ABSTRACTS, 1968-
American Society for Metals and Institute of Metals (British). Monthly. Formed by the union of: Review of Metal Literature and Metallurgical Abstracts.

Arrangement: Arranged by broad subjects subdivided into narrower topics. The subjects are classified by a decimal classification designed for metals literature. Bound cumulative volumes are published annually which bring together all the material on a given subject.

Coverage: All aspects of the science and practice of metallurgy and related fields.

Scope: International.

Locating Material: Entries are numbered consecutively through each volume. Each abstract carries a six-digit number, the first two indicating the classification number and the last four indicating the consecutive number. For example: 11-005 refers to the fifth article in the volume. The 11 indicates the subject classification.

Abstracts: Titles and abstracts are in English and the language of the article is indicated. The abstractor's initials are given or a symbol (AA) indicating author's abstract is provided. The number of references is indicated. Frequent cross references lead the user to other related abstracts.

Indexes: An author index is in each issue and an annual Cumulative Author and Subject Index is published separately. Metals Abstracts Index is a companion monthly publication issued simultaneously with Metals Abstracts but mailed separately. It is available only to subscribers of Metals Abstracts. A similar publication is Alloys Index.

Other Material: Explanatory material concerning the Index greatly facilitates its use. Also provided are Russian and Greek Alphabets, Translation Sources, and material concerning Photocopy Service. An explanation of periodical abbreviations is found just prior to the list.

Periodicals Scanned: A list appears annually. Full titles and abbreviations are given. Place of publication is given for identification purposes.

Database: Citations are available through METADEX from 1966 with monthly updates. Keywords are provided and abstracts are found after 1979.

Earth Sciences, Archaeology, and Anthropology / 75

MINERALOGICAL ABSTRACTS, 1959-
> The Mineralogical Society of Great Britain and the Mineralogical Society of America with the cooperation of 22 IMA member countries and several other countries. Quarterly. (Biennial volumes, 1959-1966, annual volumes since 1967). Previously issued as Supplement to Mineralogical Magazine and Journal of the Mineralogical Society.

Arrangement: Arranged by broad subjects subdivided into narrower topics.

Coverage: All aspects of minerals and mineralogy and related subjects, e.g., petrology, geochemistry, economic geology, gemstones, crystal structure, lunar and planetary studies, and topographical mineralogy. Book notices are a separate subheading but are included in the sequential numbering.

Scope: International.

Locating Material: Entries are numbered consecutively through the volume. The abstract number is made up of two digits representing the year, followed by the letter M/ which is followed by the abstract number. For example: 82M/0008 is the 8th abstract in 1982.

Abstracts: The abstracts are usually brief, but informative. Illustrative material and the abstractor's initials are noted. Titles and abstracts are in English and the language of the original article is indicated if it is other than English.

Indexes: An author index is found in each issue and a cumulative author and subject index is provided separately for each volume.

Other Material: A list of abstractors is found in each issue. There is also a list of abbreviations and symbols provided.

PETROLEUM ABSTRACTS, 1961-
> University of Tulsa, College of Petroleum Sciences and Engineering, Department of Information Service. Weekly.

Arrangement: Arranged by broad subjects subdivided into narrower topics. May also be divided geographically when this is indicated.

Coverage: Covers broadly the literature of petroleum and related information, including exploration, development and production, geology, geochemistry, geophysics, well logging, completion and servicing, production, storage, shipping, and supplemental technology. Includes meeting papers, patent journals, etc.

Scope: International.

Locating Material: Entries are in numerical sequence from the beginning of the publication.

Abstracts: A complete bibliographical citation is given with an indication of the language of the article and the number of references. Location of the authors is usually given.

Indexes: The Alphabetic Subject Index to Petroleum Abstracts (ASI) is divided into parts and brings together, under a single subject heading, titles of articles and patents whose abstracts may be scattered through the current awareness publication Petroleum Abstracts. The index contains the following:

    Pt. I, (1) Alphabetic Subject Index Description and Searching Procedure
          (2) Alphabetic Subject Section
    Pt. II, (1) Alphabetic Subject Index Description and Searching Procedure
          (2) Bibliographic Information
          (3) Author Index
          (4) Patent Index

There is an author index in each issue. The alphabetic subject index is issued every two months and is cumulated annually.

Other Material: Detailed explanations of use are provided in the annual indexes and a brief description is in each issue. The data are available on tape from 1965 and the subject matter is accessible using descriptors found in the Explanation and Production Thesaurus, and the Geographic Thesaurus.

Periodicals Scanned: Each issue carries a list of periodicals covered in that issue. A complete annual list of publications covered is found in the first issue of the year. This gives the full title and abbreviation.

Database: Citations are available since 1965 through TULSA. The updates are monthly and keywords are provided but not abstracts at this writing, but there are plans to include them later.

ZENTRALBLATT FÜR GEOLOGIE UND PALÄONTOLOGIE, 1950-
    E. Schweizerbart'sche Verlagsbuchhandlung. Irregular. Appears in two sections: Teil I, Allgemeine, Angewandte, Regionale und Historische Geologie; Teil II, Paläontologie. The title evolved as follows: Geologisches Zentralblatt, 1901-1942. Abteilung A, Geologie, Abteilung B, Paläontologie,

Earth Sciences, Archaeology, and Anthropology / 77

(also called Paläontologisches Zentralblatt). In 1943 merged with Neues Jahrbuch für Mineralogie, Geologie, und Paläontologie, Referate (abstracting section) to become Zentralblatt für Mineralogie, Geologie und Paläontologie, 1943-1949. In 1950 the Zentralblatt split into two sections: Zentralblatt für Geologie und Paläontologie and the Zentralblatt für Mineralogie.

Arrangement: Arranged by broad subject headings. Some issues consist of summaries of the literature on certain subjects with long bibliographies attached. For example, "Fortschritte in der regionalen Geologie Argentinienes 1963-1969" is such an article. The issues or parts of an issue are bibliographies of recent articles and books with each citation having an abstract.

Coverage: Covers all aspects of geology and paleontology.

Scope: International.

Locating Material: Entries are numbered consecutively through the year.

Abstracts: Abstracts vary in length, but the bibliographic citation is always complete. The number of illustrations is indicated and for books, price and collation information is given. Abstractor's name or his initials are provided.

Indexes: Annual author and subject indexes are supplied.

ZENTRALBLATT FÜR MINERALOGIE, 1960-
E. Schweizerbart'sche Verlagsbuchhandlung. Irregular.
Issued in two parts: Teil I, Kristallographie und Mineralogie; Teil II, Petrographie, Technische Mineralogie, Geochemie und Lagerstättenkunde. Supersedes in part Zentralblatt für Mineralogie, Geologie und Paläontologie, whose varied history is recounted in the previous entry. In 1943 superseded the Referate components of the Neues Jahrbuch für Mineralogie, Geologie, und Paläontologie.

Arrangement: Arranged by broad subjects subdivided into narrower topics.

Coverage: All aspects of mineralogy. Includes publications of societies and institutions and books, etc.

Scope: International.

Locating Material: Entries are numbered consecutively through the volume.

78 / Abstracts and Indexes

Abstracts: The abstracts are in German and in English. They vary in length and sometimes none appears. Illustrative material is indicated and the abstractor's name is provided. The language of the article is indicated.

Indexes: Each section has a cumulative author and subject index.

ZINC ABSTRACTS, 1943-
Zinc Development Association. Quarterly.

Arrangement: Arranged by six main subject headings with subsections.

Coverage: This is a survey of current world literature on the properties and uses of zinc, its alloys and compounds. Economics and statistics, metal extraction, uses and processes, pollution control, environmental aspects, health and hygiene, biochemistry, properties of materials and their measurement are among the topics covered.

Scope: International.

Locating Material: Each entry is given an alphanumeric designation. The first letter indicates the bulletin (Z, Zinc Abstracts; L, Lead Abstracts; C, Cadmium Abstracts), the next two figures stand for the year of publication and then is the abstract number which is used in the index. Also on the top line of each abstract is a classification code (CC) number (abstracts are listed in this order), a code for document type, (DT), and a language code (LC).

Abstracts: The titles are in English and in the language of the article. Illustrative material is indicated and the language of the article is noted. Patent numbers are supplied where applicable. Frequent "see references" lead the searcher to related items.

Indexes: Annual author, subject, patent, standards, and conference indexes are provided.

Other Material: A list for the classification code, code for document type, and language code is provided with the first issue of each year. Additional copies are available on request. An introduction giving an informative overview of some of the material covered is found in each issue.

Database: Citations are available through ZLC ABSTRACTS.

ENGINEERING AND TECHNOLOGY

APPLIED MECHANICS REVIEWS, 1948-
American Society of Mechanical Engineers. Monthly.

Arrangement: Arranged by broad subjects, with subtopics under each.

Coverage: Rational mechanics and mathematical methods, automatical methods, automatic control, mechanics of solids and fluids, thermal sciences, etc. are covered. (See Table of Contents on the back cover of each issue.) Books, proceedings and U.S. Government Reports are included.

Scope: International.

Locating Material: The entries are numbered consecutively through each volume. These numbers are referred to in the indexes.

Abstracts: Not really applicable as the entries are more in the nature of reviews of papers, books, etc., than abstracts of them. Titles and abstracts or reviews are in English with an indication of the original language. Reviewer's name is given at the end. Complete bibliographic citation is given including the price in the case of books. "See also" references lead the searcher to related material.

Indexes: An author index in each issue and an annual combined author and keyword index in alphabetical sequence are provided.

Other Material: Many issues have a feature article and each issue has a list of books received for reviews. A list of index terms is found in the index issues and a detailed explanation of how to use that volume. A list of reviewers is found in the index issue.

Periodicals Scanned: A journal listing is included in the annual index.

BIOENGINEERING ABSTRACTS, 1974-
Engineering Index, Inc. Monthly.

Arrangement: Arranged by subject.

Coverage: Subjects include biochemical engineering, biomechanics biomedical engineering and equipment, chemical equipment, electric power generation, fire protection, human engineering, mathematical models, prosthetics, systems science and cybernetics, telemetering, ultrasonic waves, and water supply.

Scope: International.

Locating Material: Entries are numbered consecutively through each volume.

Abstracts: Titles are in English and in the language of the article. The language is noted and the number of references is given. The senior author's affiliation is provided.

BRITISH TECHNOLOGY INDEX see CURRENT TECHNOLOGY INDEX

COMMUNICATION ABSTRACTS, 1978-
Sage Publications, Inc. Quarterly.

Arrangement: Divided into two sections: 1) journals, and 2) books and book chapters. Alphabetical arrangement by author within each section.

Coverage: A note in the preliminary pages lists "general and mass communications, advertising and marketing, broadcasting, communication theory, interpersonal and intrapersonal communications, small group and organizational communication, journalism, public opinion, speech and television" as the subjects covered.

Scope: International.

Locating Material: Entries are numbered consecutively through each volume. The index refers to these numbers.

Abstracts: The titles and abstracts are in English and the abstracts are usually quite lengthy and detailed. A list of subjects covered is found at the beginning of the abstract.

Indexes: Author and subject indexes are found in each issue. These cumulate annually.

Other Material: Brief explanatory information is given in the preliminary pages. There is limited advertising. A list titled "Briefly Noted" is provided in some issues which notes material not found elsewhere in the journal.

Periodicals Scanned: A list is provided, usually, in issue number three.

CORROSION ABSTRACTS, 1962-
National Association of Corrosion Engineers. Bimonthly.

Arrangement: Arranged by subject. There are broad areas with subtopics and numbers are assigned the broad subjects with decimals for subtopics.

Coverage: Testing, characteristic corrosion phenomena, corrosive environments, preventive measures, materials, equipment, and industries (see Table of Contents for subtopics). Proceedings, U.S. Government Reports, books, etc. are also included (see Other Material below).

Scope: International.

Locating Material: Consecutive numbers (1-92) are found down each column of each page. These numbers and the page numbers are referred to in the index.

Abstracts: Abstracts and titles are in English. The language of the article is indicated. The source of the abstract and availability of the citation are provided. Abstracts range from very brief to very detailed. The senior author's affiliation is given.

Indexes: A subject index is in each issue which cumulates annually and an annual author index is provided.

Other Material: A list of topical index headings is provided in each issue as is a list of agencies supplying abstracts (which gives abbreviations used in the entries). The first subject section, "General," includes as subtopics, reviews, bibliographies and indexes, and books. Order information for reprints is in the preliminary pages of each issue.

Periodicals Scanned: Only a list of agencies is provided as mentioned in Other Material above.

CURRENT TECHNOLOGY INDEX, 1962-
Library Association Publishing. Monthly with an annual cumulative volume. Supersedes in part the Subject Index to Periodicals. Formerly: British Technology Index, 1962-1980.

Arrangement: An alphabetical sequence of detailed subject entries and supporting cross references.

Coverage: All phases of engineering and chemical technology and manufacturing processes are covered. Also found is the chemistry and physics of man-made objects and industrial processes, instruments, and articles on the chemistry of individual substances. Does not cover industrial economics except for articles of mixed technical-economic character. In the field of management, the policy is to include only material on such physical and statistical techniques as work study, ergonomics and operational research. The biological sciences are not covered; however, borderline subjects such as the production, technology, and chemistry of food, drugs and pesticides are covered while the physiological chemical aspects are excluded. Specifically, general technology; applied science; engineering, including nuclear, mechanical, and civil engineering; chemical technology and manufacturing and technical services are among the subjects listed in the "Outline of Subject Fields...."

Scope: Covers British technical journals.

Locating Material: Searcher determines the proper subject section and scans the material.

Abstracts: Not applicable. Citations comprise subject heading, title, author, source journal, date and page location, with an indication of illustrative material and references.

Indexes: There is an author index in each issue and in the annual volume.

Periodicals Scanned: A list is found in the annual volume.

DESIGN ABSTRACT INTERNATIONAL, 1977-1981.
Pergamon Press. Ceased.

Arrangement: Arranged by subject.

Coverage: Design in general, product design, visual communications, architecture, and space planning.

Scope: International.

Locating Material: A complicated numbering system is used with various elements ending in a three-digit number. These last three numbers are consecutive through each issue and begin ".001" and continue for as many numbers as necessary. These last three digits are referred to in the indexes.

Abstracts: Titles are usually in English and in the language of the article but not always. The language of the article is indicated and English summaries are noted. A list of subjects treated in each article is found at the beginning of each entry. The name of the group who contributed the abstract is indicated by initials. Those offered by Centre Création Industrielle can be obtained in French.

Other Material: Brief explanatory material is provided.

Periodicals Scanned: A list is provided in each issue.

ELECTRICAL AND ELECTRONICS ABSTRACTS see SCIENCE AB-
STRACTS

ELECTRONICS AND COMMUNICATIONS ABSTRACTS JOURNAL, 1967-
Cambridge Scientific Abstracts. Bimonthly. Formerly: Electronics Abstracts Journal.

Arrangement: Arranged by subject.

Coverage: Among the subjects listed are: Instrumentation and measurements; automatic control; aerospace, military, medical, geoscience, safety computer and transportation electronics; amplifiers, oscillators, and waveform generators; various electronic devices and communication systems. Government reports, proceedings, books, dissertations and patents are included.

Scope: International.

Locating Material: Entries are numbered consecutively through each volume. The first two digits stand for the year and the next five are the abstract number. These begin with 00001 with each volume. The final "E" indicates the abstract journal.

Abstracts: Titles are in English and usually in the language of the article. An abbreviation indicating that language and any available summaries is given. Up to ten authors are named and the address of the senior author is provided.

Indexes: There is an author and a subject index in each issue. These cumulate annually.

Other Material: Brief explanatory information which includes a list of language abbreviations is found in the preliminary pages.

Periodicals Scanned: A list of source journals is found in the first issue of each volume.

Database: Citations are available through ELCOM from 1977. The update is every two months and abstracts are provided but not keywords.

ENGINEERING INDEX, 1884-
> Engineering Index, Inc. Monthly with annual cumulations. Previous title: <u>Descriptive Index of Current Engineering Literature.</u>

Arrangement: Arranged by main subject heading from an authority list of indexing terms. <u>Subject Headings for Engineering</u> (SHE, published separately) is provided as a guide. In some cases there are subheadings for more narrow aspects of a subject.

Coverage: Engineering, technical and applied science, and the functions of research, testing, design, construction, maintenance, production, marketing, management, consulting, and education. All aspects of engineering disciplines are covered. Papers of conferences, symposia, monographs, standards, and selected books are included.

Scope: International.

Locating Material: Each entry is assigned an abstract number. These begin with 000001 at the beginning of the monthly issues in January and at the beginning of the annual volumes. However the numbers in the monthly issues in January and at the beginning of the annual volumes bear no relationship to each other. There is a "Number Translation Index" provided.

Abstracts: Titles are in the language of the article and in English. The abstracts are in English. The senior author's affiliation is given and also the number of references. The abstracts are usually detailed giving a good résumé of the content of the article; "see" and "see also" references lead the user to alternative terms.

Indexes: Each monthly issue and the annual cumulative volume carry an author index which lists all contributing authors. The annual volume also has an author affiliation index and an abstract Number Translation Index which gives the monthly abstract number and corresponding annual abstract number. Organization names with acronyms, initials, and abbreviations are provided as is a separate list of abbreviations and acronyms.

Other Material: A detailed decription of how to use <u>The Engineering Index</u> is in Part I of the annual volume and in the monthly issues. A list of subject headings and subheadings is published separately. Additional subject headings are found in "Subject Headings Guide to Engineering Categories." An <u>Engineering Index Thesaurus</u> is published by CCM Information Corporation.

Periodicals Scanned: A list is found in the annual index. This gives CODEN, new and changed CODEN designations and cross references to CODEN titles, non-CODEN publications and conferences.

Database: Entries are available through COMPENDEX from 1969. These are updated monthly and abstracts and keywords are provided.

GAS ABSTRACTS, 1945-
Institute of Gas Technology. Monthly.

Arrangement: Arranged by broad subjects subdivided into more narrow topics.

Coverage: Among the subjects covered are planning, policy and supply; production and processing; transmission, storage and peakshaving; distribution, utilization, instrumentation and analytical methods. Government reports, academic dissertations, symposia, conference proceedings and patents are included.

Scope: International.

Locating Material: Entries are numbered consecutively through the volume beginning with 0001 with two digits preceding the consecutive number which indicate the year. Some numbers have a letter code which indicates one of the following: F - foreign; B - book; E - erratum; P - patent; R - literature review; S - second or multiple listing.

Abstracts: Titles and abstracts are in English. If the text of the article is in another language, the language is indicated. Availability of the material is indicated if other than the Institute of Gas Technology.

Indexes: Author indexes are in each issue and cumulate semi-annually and annually. Subject indexes are also provided.

Other Material: Order forms and a description of the services available are found in each issue.

Periodicals Scanned: A list is found in each issue which includes a list of associations whose publications are included.

HIGHWAY RESEARCH ABSTRACTS see TRANSPORTATION RESEARCH ABSTRACTS

HIGHWAY SAFETY LITERATURE, 1967-Sept. 1980; Fall, 1982- Transportation Research Board. Quarterly. (Published concurrently with HRIS Abstracts.)

Arrangement: Entries are in order by six-digit Transportation Research Information Service (TRIS) number. These are in ascending order but may not be consecutive.

Coverage: Accidents, injuries, fuel and fuel consumption, design, highways, impact studies, laws and law enforcement, maintenance, noise, occupants/passengers, safety devices, traffic, and vehicle design are among the subjects covered.

Scope: International. Research reports, technical papers and journal articles stored on the magnetic tape files of the Transportation Research Board are found.

Locating Material: The Fall of 1982 saw the first issue of a reactivated and refurbished Highway Safety Literature (the final issue of the previous series was published in September 1980). There are four sections: the numbered abstracts (see Arrangement above) and three indexes which refer to these numbers.

Abstracts: Titles and abstracts are in English. If the article is in another language, the title will be repeated in that language after the translated title. The language of the text is indicated at the end of the abstract. The number of references is indicated.

Indexes: A source index (which gives corporate author and addresses, and publications sources); a personal author index, including co-authors; and a retrieval term index containing subject terms from the Highway Research Information Service (HRIS) Thesaurus Term List are available.

Other Material: Explanatory material is given at the beginning of each index and the abstract section. Examples of abstracts, lists of abbreviations, availability of documents (with a list of addresses) are provided.

Database: Citations are available through HIGHWAY SAFETY LITERATURE from 1967 with monthly updates. Abstracts are provided but not keywords. This is also available as part of the TRIS system.

HRIS ABSTRACTS, 1968-
   Highway Research Information Service, Transportation Research Board. Quarterly.

Arrangement: Arranged by subject.

Coverage: Transporation administration, finance and economics, photogrammetry, highway, pavement and bridge design, materials, maintenance, safety, construction, operations and traffic control, vehicle characteristics, and human factors are covered.

Scope: Selections are compiled from computer tape records of the Highway Research Information Service. Included are contributions from various U.S. Government agencies, the Safety Research Information Service of the National Safety Council, American Society of Civil Engineers, universities, state highway departments, the United Kingdom Transport and Road Research Laboratory (Crowthorne, England), the English Language Center for the International Road Research Documentation network, and others. (See Foreword.)

Locating Material: Each entry is assigned an eight-digit code number. The first two digits indicate the subject area (as listed in the Contents). The rest are used to arrange the items within a subject area. There are usually gaps between the numbers of successive entries. These numbers are referred to in the indexes. In the subject index all entries for a particular term are listed by code number.

Abstracts: Titles are in English and in the language of the article. The language of the text if other than English is indicated at the end of the abstract. All authors' names are given. The number of references and illustrative material is noted and abstracts are generally informative. The abstractor's name is provided when known, and otherwise the source or agency supplying it is given. At the end of the abstract one finds information concerning the availability of the full text of the publication. Document order numbers for material available from the National Technical Information Service (NTIS) are given.

Indexes: Author and subject indexes (authorized term index and source index) are found in each issue. The source index is a list of journals and other sources from which the citations are taken. Mailing address and references to citations from that source are given.

Other Material: Detailed explanatory notes are provided in the foreword and pages preceding the abstracts and in each index. A list of abbreviations is given.

Periodicals Scanned: A list is in each issue (see Source Index).

Database: Citations are available through TRIS from 1970 and are updated monthly. Abstracts and keywords are provided. In addition to HRIS Abstracts, RRIS Bulletin, MRIS Abstracts and other references are also included in the database.

OFFSHORE ABSTRACTS, 1974-
Offshore Information Literature. Bimonthly.

Arrangement: Arranged by subject.

Coverage: Arctic environment, buoys and mooring systems, cables, cathodic protection, corrosion, economic surveys exploration, general surveys, legislation, marine mining, metallurgical research, oceanography, pipes and pipelines, production, pumps, structural engineering, underwater equipment, and welding are the subjects listed in the Contents. Books, conferences, and reports are included.

Scope: International.

Locating Material: Entries are numbered consecutively through the volumes. The first two digits indicate the year.

Abstracts: Titles and abstracts are in English. Illustrative material and the number of references are indicated. Occasionally no abstract is provided.

Other Material: Brief explanatory material is furnished. The introduction gives the subject matter included in some of the entries.

ROAD ABSTRACTS, 1934-1968.
Technical Information and Library Group, Road Research Laboratory. Ceased. Volumes 1-4 issued as Supplement to the Journal of the Institution of Municipal and County Engineers, Volumes 5-16 issued by the Institution of Municipal and County Engineers in cooperation with the Institution of Civil Engineers; Volumes 17-35 issued by the Road Research Laboratory.

Arrangement: Arranged by subject. Twenty-four subject headings are listed in the Table of Contents.

Coverage: Covers all aspects of roads and traffic including construction, bridges, pertinent geological, climatological, hydrological and economic material, road layout, parking, accidents and safety measures, street lighting, road users, and other relevant material.

Scope: International.

Locating Material: Citations are in numerical sequence through the volume.

Abstracts: A complete bibliographic citation is given, with abstracts ranging from one line to a half page. Illustrative material is indicated as is the language of the article, and sometimes the price of a monograph, and the IRRD (International Road Research Documentation) serial number is given. If the summary is the author's, this is indicated.

Indexes: An annual subject and name index is provided.

Periodicals: An annual list of journals is furnished which gives abbreviations used, the name of the publisher and his address. There is a further list of titles by country of origin provided.

SAFETY SCIENCE ABSTRACTS JOURNAL, 1978-
Cambridge Scientific Abstracts.   Bimonthly.

Arrangement: Arranged by broad subjects subdivided into narrower topics.

Coverage: The broad subjects include industrial and occupational safety, transportation safety, aviation and aerospace safety, environmental and ecological safety, and medical safety. An introductory note states that: "Areas of coverage include pollution, fire, waste disposal systems, radiation, drug dosages, pesticides, epidemics, and ... other events or phenomena which ... threaten mankind, his environment, or the technology upon which he depends."

Scope: International.

Locating Material: Entries are numbered consecutively.

Abstracts: Titles and abstracts are in English. The language of the article is noted and the senior author's affiliation is given. "See references" at the beginning of subject sections lead the researcher to related items.

Indexes: Subject and author indexes are found in each issue which cumulate annually.

Other Material: Very brief explanatory material is provided and a list of the publisher's related abstract journals (with descriptions) is found.

Periodicals Scanned: A Source Index which contains a list of the periodicals regularly scanned is contained in issue number one of each volume. Publishers' addresses are included. A complete list is available on request.

Database: Citations are available through Safety Science Abstracts Journal from 1975. The update is every two months.

TRANSPORTATION RESEARCH ABSTRACTS, 1931-1975.
Highway Research Board, National Research Council. Ceased. Number one (May 1931) and three (February 1932) have title: Research Abstracts. Formerly: Highway Research Abstracts (1931-June 1974).

Arrangement: There is a random arrangement.

Coverage: Highway transportation and related subjects including accidents, air pollution, bridges, cement, drainage, mass transportation, planning, design, vibration, etc. are found (see Table of Contents).

Scope: International.

Locating Material: A list of the titles of articles included in each issue is found on the front and back cover of each issue with reference to page number. A subject index gives titles under the subject heading with reference to volume, issue, and page number; e.g., a subject entry followed by 40 (8) 12, Aug., 1970 means that the abstract is in Volume 40, no. 8, page 12, August 1970.

Abstracts: Titles and abstracts are in English. If the article is in another language, the language is specified. Availability of an article is given either by the address of the publisher or the author, report numbers and the price in the case of government reports. Illustrative material is noted. Abstracts are usually concise but informative.

Indexes: An annual subject index since 1962 and a cumulative "Index to Highway Research Abstracts 1931-1961" are available. The annual subject index is arranged by subject with the title of the citation being listed chronologically under the subject. A Table of Contents of the subject headings precedes this index. The December issue ("Tentative Program and Abstracts of Papers to be Presented at the Highway Research Board Annual Meeting" until 1974 and simply "Tentative Program of the Annual Meeting," 1974-1975) is not indexed. The index contains an alphabetical list of subjects.

Other Material: The December issue contains information about the annual meeting including a tentative program, a schedule of meetings, and a list of participants.

URBAN MASS TRANSPORTATION ABSTRACTS, 1972-
Transit Research Information Center. Bimonthly.

Arrangement: Arranged by subject.

Coverage: Among the topics covered are: conventional transportation services, center city/traffic restraints, energy and environment, fares, pricing and service innovations, financing, land use, nonurban and low density area transportation, paratransit systems and services, planning, policy, and program development, political processes and legal affairs, safety and product qualification and security, socioeconomics, technology development, transportation of the disadvantaged and special user groups, transit management, transportation productivity and efficiency, and urban goods movement.

Scope: Contains abstracts of "research, development and demonstrations, technical studies, and university research projects sponsored by the Urban Mass Transportation Administration."

Locating Material: Each subject section is numbered and entries are arranged alphabetically by title within these sections. Reports in the abstract section are available from NTIS.

Abstracts: Each entry gives the title, personal or corporate author, the source of funding, date of publication, NTIS order number and a price code. The abstracts are usually quite lengthy and detailed.

Indexes: A cumulative retrieval term index appears at the end of every third issue.

Other Material: There is a list of technical studies acquisitions (resulting from Section 8/9 of the Urban Mass Transportation Act of 1964) arranged by state. These are not available from NTIS.

Database: Citations are available through TRIS from 1968 and are updated monthly. Keywords and abstracts are available.

ENERGY AND ENVIRONMENT

AIR POLLUTION ABSTRACTS, 1932-1972.
Warren Springs Laboratory (Stevenage, England). Ceased.
Formerly: Atmospheric Pollution Bulletin.

Arrangement: Arranged in seven subject categories.

Coverage: Emissions and sources, identification and measurement, distribution, effects, administration, method and equipment for abatement (see Contents list).

Scope: International.

Locating Material: Searcher finds the proper subject category and scans the material.

Abstracts: Titles and abstracts are in English. The language of the article and summaries in other language are indicated and the number of references is given. If the abstract is from another source, that source is cited. An asterisk indicates that the complete article or English summary has been reproduced and other information in English cannot be obtained from the literature.

Other Material: A list of sources of abstracts other than the Warren Springs Laboratory Library Staff is given in the preliminary pages of each issue.

AIR POLLUTION ABSTRACTS, 1970-1976.
Environmental Protection Agency (U.S. Government Printing Office). Ceased. Previously issued by the U.S. National Air Pollution Control Administration as NAPCA Abstract Bulletin.

Arrangement: Arranged by subject. Entries are numbered consecutively from Volume 2, no. 5. The APTIC (Air Pollution Technical Information Center) accession number is also provided. The abstract numbers are referred to in the indexes.

Coverage: Emission sources; control measurement methods; air quality measurement; atmospheric interaction; effects on human health, plants, and livestock; economic, social, legal, and administrative aspects. (See list of subject fields and scope notes.) Includes journals, patents, reports, dissertations, preprints, etc.

Scope: International, but material covered is from "technical literature recently accessioned by the APTIC."

Locating Material: An abstract (accession) number is assigned each entry. These numbers are referred to in the indexes. There is some variation in form over the years, e.g., in the earlier issues reference is also made to "field" or class number and the searcher turns to the proper section to discover a given numbered abstract.

Abstracts: The abstracts are in English. Titles are in English and the language of the article. The language of the article is noted and the number of references is given.

Indexes: Author and subject indexes are found in each issue which cumulate twice a year.

Other Material: Order information is found in the preliminary pages of each issue.

Periodicals Scanned: A list is found in one issue of each volume.

Database: Entries are available through AIR POLLUTION TECHNICAL INFORMATION CENTER (APTIC). The database has material from 1966-1978.

ECOLOGY ABSTRACTS, 1975-
Cambridge Scientific Abstracts. Monthly. Formerly: Applied Ecology Abstracts.

Arrangement: Arranged by broad subjects subdivided into narrow topics.

Coverage: Ecosystems in general and coastal, aquatic, and terrestrial ecosystems are covered. Types of land (woodland, grassland, etc.), various kinds of animals and plants and their interactions, human ecology, management, conservation, and reclamation are also included.

Scope: International.

Locating Material: Entries are numbered consecutively through each volume. There are three parts to these numbers. First a number for the abstract, a code letter identifying the journal, and finally the volume number.

Abstracts: Titles are in English and in the language of the article. The language and summaries in other languages are indicated. The abstracts are in English.

Indexes: There are author and subject indexes in each issue which cumulate annually. Since 1980, abstracts are provided for all Man and Biosphere (MAB) publications (this is a program of Unesco); these are found under MAB in the subject index.

Other Material: Descriptive information is provided in the preliminary pages.

Periodicals Scanned: A list of periodicals searched is available on request. A separate list covering the Cambridge Scientific Abstracts life science journals is available for sale.

Database: Citations are available from 1978 through IRL LIFE SCIENCES COLLECTION and are updated monthly. Keywords and abstracts are provided.

ENERGY: A CONTINUING BIBLIOGRAPHY WITH INDEXES see SCIENTIFIC AND TECHNICAL AEROSPACE REPORTS.

ENERGY ABSTRACTS FOR POLICY ANALYSIS, 1975-
U.S. Department of Energy, Technical Information Center (U.S. Government Printing Office). Monthly.

Arrangement: Arranged by subject.

Coverage: The coverage is intended to emphasize "programmatic efforts; policy, legislative and regulatory aspects; social, economic, and environmental impacts; regional and sectoral analyses, institution factors, etc." Among the subjects included are economics and sociology, natural resources, nuclear energy, conservation, fossil fuels, consumption and utilization, unconventional sources, and power generation. A list is found under Subject Contents in the front of each issue. (Not all issues will include all subjects, and only those found in a particular issue will be listed there.)

Scope: Substantive articles from Congressional Committee prints, federal, state, and local agency reports, periodicals, conference papers, books and organizations or institution reports are found. Entries are limited to nontechnical or quasi-technical material.

Locating Material: Entries are numbered consecutively through each volume. These numbers plus the volume number are referred to in the indexes.

Abstracts: Titles and abstracts are in English. The language of articles not in English is indicated. The abstracts vary in length from one line to very detailed analyses. An entry for a government report includes the report number, contract number, availability, and corporate source. Those for journal articles give complete citations. All personal authors are named.

Indexes: Corporate authors, personal authors, subject and report number indexes are included in each issue. These cumulate annually. The corporate author index gives the institution, the title and a report number; all personal authors are indexed, but only initials are used for given names. Co-authors are referred to the senior author. Titles are given in this index, and in the subject index. The report number index gives the availability and a price code. All indexes refer to the volume and abstract number.

Other Material: Introductory material provides explanations of the entries, the criteria for inclusion and the indexes.

Database: Entries are available through DOE/RECON database EDB from 1975. These are updated every two months and abstracts and keywords are provided. The database also includes Energy Research Abstracts.

ENERGY INDEX see ENERGY INFORMATION ABSTRACTS

ENERGY INFORMATION ABSTRACTS, 1976-
Environment Information Center. Monthly. The Annual Cumulation, Energy Information Abstracts Annual includes all the abstracts from the monthly journal. The annual index, The Energy Index, can be used with the monthly issues, the annual cumulation or independently.

Arrangement: Arranged by subject. There are twenty-one subject categories.

Coverage: Energy information from government agencies, journals, institutions, and corporations. Among the subjects included are policy and planning, petroleum, gas and coal resources, fuel processing, transportation and storage, electric power storage and transmission, thermonuclear power, and industrial and residential consumption.

Scope: International.

Locating Material: Abstracts are consecutively numbered through each volume. The number is preceded by two digits which

stand for the year and begin the volume with 20001. Energy
Information Abstracts Annual is a reproduction of each monthly issue in the order in which it originally appeared. The
volume number and issue number are noted at the bottom of
each page. Energy Index refers to the abstract number.

Abstracts: Titles and abstracts are in English. Senior author's
affiliation is given if possible and illustrative material and
the number of references is noted. "See references" at the
beginning of each subject section lead the searcher to alternate material. Some abstracts are marked by an asterisk
(*) and can be ordered in hard copy or microfiche from EIC.

Indexes: The Energy Index is compiled from the monthly issues.
Access is provided by subject, Standard Industrial Classification (SIC) code, geographic location, source, and author.
A keyword list is provided which guides the user to the
proper subject term. The title and complete citation is provided so a searcher not interested in the abstract could go
directly to the article or report.

Other Material: Explanations of how to use the publication are
provided and a description of EIC products is given. A list
of abbreviations used is in the annual index. Articles, a list
of conferences, a list of books, statistics, legislation, and
an "issue alert" give overviews in the Annual Index and Abstracts volume of significant information appearing during the year.

Periodicals Scanned: A list is provided in the Annual Index and
Abstracts volume.

Database: Citations are provided from 1971 through ENERGY LINE.
There are monthly updates and abstracts and keywords are
provided.

ENERGY RESEARCH ABSTRACTS, 1976-
U.S. Department of Energy, Technical Information Center
(U.S. Government Printing Office). Semimonthly. Formerly:
ERDA Energy Research Abstracts.

Arrangement: Arranged by broad subject headings subdivided into
narrower categories.

Coverage: Publications of the Department of Energy and material
in report form by federal and state government agencies,
foreign governments and institutions; journal articles, conference papers, patents, theses, and monographs are included if they originated in the Energy Department. Subjects
include coal, petroleum, natural gas, nuclear and fusion
fuels, hydro, solar and wind energy, energy storage and
conversion, chemistry, engineering and various aspects of
environmental sciences.

Scope: International material in report form. Domestic publications as mentioned above.

Locating Material: Abstracts are numbered consecutively through each volume. In the indexes, volume and abstract numbers are given.

Abstracts: Sometimes quite brief but usually very detailed. For reports, contract and report numbers, availability, and corporate source are given and for journal articles, complete citations are provided. All personal authors are named. "See also" references at the beginning of a new subject section lead the searcher to related material.

Indexes: Corporate source, personal author, subject, contract and report number indexes are provided in each issue. These cumulate semiannually and annually and are available in printed form through Volume 6. From Volume 7 they are available also on microfiche. All index entries refer to the volume and abstract number. Titles are given except in the contract number index (which also provides the report numbers) and the report number index which gives information on where a report can be obtained and a price code.

Other Material: A numerical and an alphabetical list of the subject contents is provided. An explanation of the price code and an explanation of the abbreviations used in the availability column is provided in the back of each issue. A brief list of institutions holding energy report collections is provided on the inside back cover of each issue.

Database: Information is available through EDB from 1975. This is updated every two months and abstracts and keywords are provided. There are other publications besides Energy Research Abstracts derived from this database which is part of DOE/RECON.

ENVIRONMENT ABSTRACTS, 1971-
Environment Information Center, Inc. Monthly, with an annual cumulation, Environment Abstracts Annual. Volumes 1-3 were titled Environment Information Access and were issued semimonthly.

Arrangement: Twenty-one categories are listed as classifications which offer a very full coverage of the environment. These include air, water, and noise pollution; land use and misuse; nonrenewable resources; environmental education and design; urban ecology; population planning; energy; wildlife, and weather modification, Both print and nonprint media are included as are journal and newspaper articles, important radio and television programming, films, filmstrips. Conference reports, entries from the Federal Register, and patents are found. The List of Classification describes what is included in each of the twenty-one subjects.

Scope: International.

Locating Material: Entries are listed by a seven-digit accession number through the volume. The first two digits indicate the year and the rest is a sequential number (81-00001 is the first entry in 1981). These numbers are referred to in the indexes. "See references" at the beginning of each subject section guide the user to alternate or related terms.

Abstracts: Titles and abstracts are in English. The type of information, i.e., research summary, commentary, feature article, editorial, news analysis, etc., is given. Major reports receive extended abstracts while others will be more brief. Following the abstract and enclosed in parentheses is a notation of the number of references and illustrative material. An asterisk before the accession number indicates that it is available on microfiche.

Indexes: Author, subject, and industry indexes are in each issue. An annual cumulation, The Environment Index is published separately (as a companion to Environmental Abstracts Annual). In this, there are five separate indexes: Subject, Standard Industrial Classification (SIC) Code, Geography, Author, and Source (since 1981). For the subject index, several thousand keywords are employed.

Other Material: In The Environment Index one finds a review of the literature of the year, articles on national and international aspects of the environment, legislation, a list of environmental impact statements, a directory of federal, regional, and state offices and agencies, non-governmental organizations, and bibliographic databases, and a list of conferences, books, and films. Some of these features appear in the monthly issues.

Periodicals Scanned: A list of periodicals appears in the annual index.

Database: Entries are available from 1971 through ENVIROLINE and are updated monthly. Abstracts and keywords are provided.

ENVIRONMENT INDEX see ENVIRONMENT ABSTRACTS

ENVIRONMENT INFORMATION ACCESS see ENVIRONMENT ABSTRACTS

ENVIRONMENTAL QUALITY ABSTRACTS, 1975-
Data Courier, Inc. Quarterly.

Arrangement: Arranged by subject.

Coverage: In the "statement of purpose" it is said that this journal "...organizes, summarizes and indexes worldwide literature in the environmental sciences. It is designed to assist students and educators at the secondary school and undergraduate college levels in keeping current on latest scientific developments and in conducting special research projects." Among the topics listed in the contents are public policy, population and health, conservation and endangered species, environmental contamination, resources and recycling, and energy resources.

Scope: International.

Locating Material: Entries are numbered consecutively through each volume. The year and volume number are first, then a hyphen and the sequential number.

Abstracts: Abstracts and titles are in English. The abstracts are usually long and give good information.

Indexes: An author index and a keyword index are provided in each issue.

Other Material: Detailed explanatory information is given on finding material. A list of abbreviations and symbols is provided; articles and book reviews are to be found in a separate section.

Periodicals Scanned: A list of publications regularly cited is found in each issue. There is also a list of source journals with addresses provided.

FUEL AND ENERGY ABSTRACTS, 1960-
Butterworth Scientific, Ltd. (for Institute of Energy). Bi-monthly.

Arrangement: Arranged by 27 broad subject headings. Alphabetical by title in each section.

Coverage: Natural and derived solid fuels, natural and derived liquid fuel, natural and derived gaseous fuel, by-products related to fuels, nuclear energy, electric power generation, alternative energy sources, combustion, space heating and cooling systems, environment and energy are among the topics listed in the Contents.

Scope: International.

Locating Material: Entries are alphabetically arranged by title in each subject section. A code number follows each entry which indicates the year, the section number, and the entry number in the section. There are numerals to indicate subsections where they occur. (83-01/02-0001 means the first entry in section 2 of the first issue of 1983.)

Abstracts: Titles and abstracts are in English. If the article is in another language, the language is indicated. Abstracts vary in length from quite long to very brief. The source of the abstract is provided.

Other Material: A list of abbreviations for abstract sources is provided. There are brief explanatory notes.

HYDATA, 1965-
American Water Resources Association. Monthly.

Arrangement: Arranged alphabetically by the name of the journal.

Coverage: A wide range of journals is included relating to hydrology, water resources, coastal engineering, water power, water research, and technology.

Scope: International.

Locating Material: The Table of Contents guides the user to the correct section of each issue and then the journal titles are arranged alphabetically.

Abstracts: Not applicable. This publication reproduces the Table of Contents of journals likely to have articles concerning hydrology and water resources. There are also selected titles of nonperiodical literature, and Tables of Contents of nonperiodical literature, Water Resources Review, and New Publications of the U.S. Geological Survey.

Indexes: An annual index was provided for a few years (1967-1970?) by Hydor: A Bibliography and Index of the Articles, Monographs, and Reports Listed in Hydata.

Other Material: A membership plan for the American Water Resources Association and a list of their district directors and section officers is given. The kind of material found has changed over the years and has included illustrations from articles or monographs, a newsletter, a list of meetings, and presently, a section called "Water Resources Library."

Periodicals Scanned: A list is found in each issue.

LAND USE PLANNING ABSTRACTS, 1974-1979.
Environmental Information Center. Ceased.

Arrangement: Arranged by subject.

Coverage: Air pollution, chemical and biological contamination, energy, environmental education, environmental design and urban ecology, food and drugs, land use and misuse, noise pollution, nonrenewable resources, oceans and estuaries, population planning and control, radiological contamination, renewable resources of land and water, solid waste, transportation, water pollution, weather modification and geophysical change, and wildlife are listed as the review classifications.

Scope: International.

Locating Material: Each subject section is numbered and entries are numbered consecutively from the beginning of the publication. The first two digits indicate the year. A list of keywords and subject terms is provided which leads the user to the proper section and abstract number.

Abstracts: Titles and abstracts are in English. The senior author's affiliation is given (only the first two authors are named; for others, "et al." is given). Illustrative material is noted and the number of references is provided.

Indexes: Subject terms (keyword list) and subject index; Standard Industrial Classification (SIC) Code terms (keyword list) and SIC code index; geographical terms (keyword list) and geography index; and author index, are provided with each annual volume.

Other Material: Articles; a directory of databases, officials, and periodicals; legislation; a list of conferences; books and films; and a section of statistics are found.

Periodicals Scanned: A list is provided in each annual volume.

PESTICIDES ABSTRACTS, Volume 7, Number 1, 1974-1981.
U.S. Environmental Protection Agency (U.S. Government Printing Office) Ceased. Title varies: Health Aspects of Pesticides Abstracts Bulletin, 1969-1974.

Arrangement: Arranged by subject.

Coverage: General material which includes news items, letters, and editorials appearing in scientific and technical journals. Specifically listed are: Phytoxicity studies, monitoring of pesticides in humans or nonexperimental animals as well as water, air, soil, food, feed and consumer products; pesticide poisoning in man and nonexperimental animals, its treatment, morbidity and mortality; safety measures and decontamination; legislation and regulation; toxicity studies and the effects on cells, organs and systems; metabolic studies, biochemistry and interaction, antidotes and mode of action. A section on analysis covers preparation of samples, analytical methods, and instrumentation. The Table of Contents gives an explanation of the broad topics.

Scope: International.

Locating Material: The entries are numbered consecutively through each volume. The abstract number begins with two digits which indicate the year and the rest is a sequential number beginning with 0001. These numbers are referred to in the indexes.

Abstracts: All authors are listed and the affiliation of the senior author is found. Titles are given in the language of the article and in English. The language is noted as are summaries in other languages. The number of references is indicated. "See also" references at the end of each subject section lead the user to related items.

Indexes: Produced quarterly and cumulated annually. There are two subject indexes (concept and compound) and two author indexes (personal and corporate). A list of the broader terms used in the concept index is provided. Compounds are indexed by their common names.

Other Material: A description of the subjects covered is found in the Table of Contents.

Periodicals Scanned: A list appears annually in the January issue.

PESTICIDES DOCUMENTATION BULLETIN, 1965-1969.
U.S. Department of Agriculture, National Agricultural Library (U.S. Government Printing Office). Ceased.

Arrangement: Arranged by subject. Entries are alphabetical by author under subject.

Coverage: Entomology; crop, livestock and commodity protection; environmental contamination; toxicology; chemistry; etc. are covered. (See Table of Contents.)

Scope: International.

Locating Material: Each entry is given an accession number. The last two digits represent the Bulletin's issue year. The first part of the number is assigned serially and follows consecutively through each volume. These numbers are referred to in the indexes.

Abstracts: The titles and very brief abstracts or list of subjects covered in an article are in English. An abbreviation is given which indicates the language of the article if it is other than English. All personal authors are given and also the number of references and the National Agricultural Library call numbers are noted. Abstracts are usually quite brief. Note is made of summaries in language other than that of the article.

Indexes: Subject and biographical indexes (names of first author arranged alphabetically with affiliation and accession number) are found. Author index, and organizational (names of corporate author or sponsoring organizations arranged alphabetically with accession numbers of pertinent citations) index are in each issue. The indexes are cumulated biannually.

Other Material: A description of the Bulletin and availability of references cited is in the preliminary pages of each issue and the cumulative indexes.

POLLUTION ABSTRACTS, 1970-
Cambridge Scientific Abstracts. Bimonthly. (Publisher varies.)

Arrangement: Arranged by subject, i.e., the element being polluted (land, water, etc.), then other related subjects.

Coverage: Pollution control and research concerning land, water, air, and noise. Sewage and wastewater treatment, waste management, toxicology and health, radiation and environmental action are also covered. Conference proceedings, books, government reports, and documents with limited circulation are also included.

Scope: International.

Locating Material: Entries are numbered consecutively through each volume with the first two digits representing the year. These numbers are referred to in the indexes.

Abstracts: Information in the preliminary pages of each issue states that the abstracts "...give the basic information reported in the original material, including conclusions and methods." Titles and abstracts are in English and the language of the original article is noted by an abbreviation if it is other than English. The senior author's affiliation is given. A list of related subjects and abstract numbers is provided at the beginning of each section.

Indexes: Author and subject indexes are in each issue which cumulate annually. In the subject index, each abstract is given up to twelve terms which indicate the main subjects covered. These terms are permuted to allow the user several access points.

Other Material: A description of how to use the journal, a list of acronyms and of abbreviations, and a "Calendar of Events," is found in each issue. A "Controlled Vocabulary" from which the subject terms are taken is included in the cumulative indexes.

Periodicals Scanned: A master list is included in the cumulative index at intervals (see 1980 Annual Cumulative Index). Additions to that list are found in each issue.

Database: Citations are available through POLLUTION ABSTRACTS from 1970. They are updated every two months; abstracts are provided from 1978 and keywords are supplied.

SELECTED WATER RESOURCES ABSTRACTS: 1968-
Office of Water Research and Technology, U.S. Department of the Interior. (U.S. Government Printing Office) Monthly. (Semimonthly through 1981.)

Arrangement: Arranged in ten major subject areas, each designated by a two-digit number ranging from 01-10. These in turn are divided into sixty smaller subject areas, designated by letters of the alphabet. For example: Section 1 is "Nature of Water"; Section 1-a is "Properties." Further, citations are arranged under accession numbers consisting of the letter W (which identified the item as belonging to WRSIC) followed by the last two digits of the year of the issue, followed, in turn, by a five-digit number representing the sequential accession number. For example, W72-00001 refers to the first citation acquired in 1972.

Coverage: Water-related aspects of the life, physical, and social sciences as well as relevant related material in engineering and legal aspects of water, and the characteristics, conservation, control, and management of water. Covers serial publications and current and earlier pertinent monographs. Supplementary documentation is obtained from Biological Abstracts and BioResearch Index which provide relevant references.

Scope: International.

Locating Material: The index reference is to the location of the citation in its specific subject area. Its accession number is also given to further pinpoint its location. For example: in locating a citation in the monthly issues' indexes, 06B with an accession number W72-09005 means that the user must refer to the subject section 6B, then find the numerical sequence number 09005. In the cumulative indexes the numbers are expanded as follows: 1421 5 F with an accession number of W81-05338 means that the user must refer to Volume 14, Number 21, subject section 5F, then to number 05338 in that subject section.

Abstracts: Full bibliographic information is provided and the abstracts give detailed information. Illustrative material and the number of references are indicated. The location of the research is given and a list of subject headings (descriptors) which are taken from the Water Resources Thesaurus are provided.

Indexes: Monthly author, subject, organizations participating, and accession number indexes are found. Annual author, organization, subject, and accession number indexes (all of which list the title of the paper except the accession number index) are provided.

Other Material: A list of Centers of Competence is in the annual index. A brief explanation of the publication is provided in the preface of each issue. A thumb index aids in the use of this work and a price code is provided on the back of each issue. A table of accession numbers correlated to issue numbers facilitates the use of the annual indexes.

Database: Citations are available through WATER RESOURCES ABSTRACTS (WRA) from 1968.

WATER POLLUTION ABSTRACTS, 1927-1973.
    Department of the Environment. (Her Majesty's Stationery Office). Ceased. Formerly: Summary of Current Literature, 1927-1929; Water Pollution Research, Summary of Current Literature, 1930-1948.

Arrangement: Arranged by subject.

Coverage: Conservation, analysis, and examination of water and waste, sewage, trade waste waters, and effects of pollution are covered. Books, proceedings, and symposia are included.

Scope: International.

Locating Material: The entries are numbered consecutively through each volume. These numbers are referred to in the indexes.

Abstracts: Titles and abstracts are in English. The language of articles and summaries in other languages are noted and illustrative material is indicated. Abstracts are usually brief but generally quite informative. Sources of abstracts are given if they are other than that of the Department of Environment, Water Pollution Research Staff. If there is another source for the article or the same information it contains, a reference is made at the end of the abstract. Periodical abbreviations are those given in World List of Scientific Periodicals.

Indexes: Annual author and subject indexes are published separately.

BIOLOGICAL SCIENCES

AMINO ACIDS, PEPTIDES AND PROTEINS see BIOCHEMISTRY ABSTRACTS

ANIMAL BEHAVIOR ABSTRACTS, 1973-
Cambridge Scientific Abstracts. Quarterly. Formerly: Behavioural Biology Abstracts, 1973-1974.

Arrangement: Arranged by subject.

Coverage: Subjects include aggression, dominance, sexual and social behavior, avoidance and discrimination learning, orientation, navigation and migration, locomotion, visual and other external stimulation, motivation, rhythms, sleep, electrical and chemical brain stimulation, and methodology. The foreword states: "Papers devoted to subjects which border on behaviour, ranging from neurophysiology to ecology and from genetics to social anthropology are covered ... also, the important journals dealing with the biology of particular taxonomic groups."

Scope: International.

Locating Material: Entries are numbered consecutively through each volume. There are three parts to these numbers: first a number for the abstract, then a code letter identifying the journal, and finally the volume number.

Abstracts: Titles are in English and in the language of the article. The language and summaries in other languages are indicated. The first ten authors' names are given and the address of the senior author is provided.

Indexes: Author and subject indexes are provided in each issue. These cumulate annually.

Other Material: There is a section explaining how to use the journal, some descriptive information is provided, and a list of abbreviations is given.

Periodicals Scanned: A list of periodicals is provided on request. A separate list covering Cambridge Scientific Abstracts life science abstract journals is available for sale.

Database: Citations are available from 1978 through IRL LIFE SCIENCES COLLECTION and are updated monthly. Keywords and abstracts are provided.

APICULTURAL ABSTRACTS, 1950-
International Bee Research Association. Quarterly.

Arrangement: Arranged in classified order by Universal Decimal Classification (UDC); many subdivisions occur which are more finely divided by the use of capital letters. UDC order is followed within each main subject group. The following example is listed in Notes on Apicultural Abstracts: "... 638.15 represents honeybee diseases and enemies; within this, 638.153 is diseases of adult honeybees, which is itself divided into respiratory diseases, 638.153.2, and digestive diseases, 638.153.3...." The capital letter A is used to further subdivide amoeba disease from nosema disease, indicated by the capital letter B.

Coverage: Covers comprehensively the literature or research and technical developments concerning all bees and beekeeping. All aspects of Apis Mellifera (honeybee) and other Apis species are found. Includes Apoidea in general, especially foraging and social behavior and pollinating activities; does not deal exhaustively with taxonomy. Relevant material on the social aspects of other insects and general aspects of pollination are covered as well as the economic use of social and solitary bees for crop pollination. There is full coverage of developments in techniques and equipment for beekeeping and for processing hive products. Primary publications or original research is abstracted.

Scope: International.

Locating Material: Entries are numbered consecutively through each volume. The number also includes the last two digits of the year and may carry the letter "L" indicating there is no abstract or author's address provided (24L/72 is the twenty-fourth citation in 1972 with no abstract or author's address found). Reference to the citation in the annual index does not include the year designation.

Abstracts: Titles are in English and in the language of the article. Abstracts are in English. The language of the article and of available summaries is indicated. UDC numbers are given for each citation. The abstractor's name is provided. At the right of the citation is the letter B, indicating that the publication abstracted is in the Bee Research Association Library and/or the letter E which means that an English translation is in the library.

Indexes: Annual author and subject indexes are provided. Index to Apicultural Abstracts, 1950-1972 and Index to Apicultural Abstracts, 1973-1980 (COM microfiche) is available separately.

Other Material: Explanation of the subject index is provided as is a discussion of computer indexing and services. The Table of Contents gives a detailed list of the subjects covered. "Notes on Apicultural Abstracts," found in each issue, discusses the abstracts and their services. A full English Alphabetical Subject Index (EASI) is available separately. Information leaflets are available which describe materials available and provide other information. A list of abstractors follows the author index. Pollination of Seed Crops, ed. by Eva Crane is a summary of research reported in Apicultural Abstracts, 1959-1971.

Periodicals Scanned: List L9 titled: "The Journals Yielding Most Papers for Apicultural Abstracts: 250-300 Journal Titles Based on an Analysis for 1961-1972" is available by purchase.

Database: Citations are available through CAB ABSTRACTS from 1973. Monthly updates and keywords and abstracts are provided.

AQUATIC SCIENCES AND FISHERIES ABSTRACTS, 1971-
Cambridge Scientific Abstracts. Monthly. Formed by the union of: Current Bibliography for Aquatic Sciences and Fisheries, 1958-1971 (April 1958 issue under title Current Bibliography for Fisheries Science) and Aquatic Biology Abstracts, 1969-1971. In 1978 divided into two parts: Pt. 1, Biological Sciences and Living Resources, and Pt. 2, Ocean Technology, Policy and Nonliving Resources.

Arrangement: Arranged by broad subject areas subdivided into smaller topics.

Coverage: A broad spectrum of information on oceanography, limnology, biology, ecology, pollution, diseases, and information on fisheries, ocean policy, technology, physical and chemical oceanography is included. In addition to journals, books and proceedings are checked for information. Abstracts are put in the most relevant section with cross references to related sections. Animals whose life cycles are spent in both freshwater and marine waters are classified as brackish.

Scope: International.

Locating Material: Entries are numbered consecutively through the volume. These numbers have three parts: the first is the sequential number of the abstract, second is a code letter which identifies the abstract journal, and the third is the volume number.

Abstracts: Titles are in English and in the language of the article. Abstracts are in English. Titles in Cyrillic or other non-Roman alphabets are transliterated. The language of the article and of any available summaries is indicated. The address of the senior author and the number of references are noted.

Indexes: There are monthly and annual cumulative author, taxonomic, geographic, and subject indexes. The taxonomic index is alphabetical by scientific name, the common names being used only when scientific names are not available. The names are taken from the papers themselves and other reference sources, including mainly, the "National Oceanographic Data Center Taxonomic Code." Species names appear as in the original text. The geographic index is divided into sections: land areas, sea areas, inland waters, and coastal areas. Land areas are alphabetical by the English name of the country; sea areas (including inland seas and interterritorial lake systems) are indexed under the main water areas. Coastal regions are indexed under sea areas and qualified by the land area name. Descriptors in the subject index are from a "Thesaurus of Terms for Aquatic Sciences and Fisheries" compiled by the FAO (FAO Fisheries Circular no. 344, currently being revised).

Other Material: A list of abbreviations used in the text and for languages is provided. The Table of Contents in each issue serves as an outline of the classification. The Preface of each issue and of the annual indexes provides an introduction to the user. Bibliographic captions, abstracts and index entries for both Parts 1 and 2 have been available since 1978 in tape format (an experimental tape carrying index terms for 1975 and bibliographic data for 1976 and 1977 is available from FAO).

Database: Citations are available since 1978 through ASFA. There are monthly updates and keywords are provided but no abstracts.

BEHAVIOURAL BIOLOGY ABSTRACTS see ANIMAL BEHAVIOR ABSTRACTS

BIBLIOGRAPHY OF BIOETHICS, 1975-
Gale Research Company. Annual.

Arrangement: Arranged alphabetically by broad subjects relating to bioethics; some have subtopics for various aspects of a given subject. Within each subject the arrangement is alphabetical by author.

Coverage: Abortion and relevant problems such as legal and religious aspects, financial support, and public and professional attitudes; artificial insemination, in vitro fertilization, cloning, genetic screening, allowing to die, human experimentation, organ donation and transplantation, patients' rights, sociobiology, recombinant DNA research, and problems of war are covered. A complete list is provided by the Table of Contents of each volume (which may change from year to year). The subject headings in Volume 7 (1981) are different and simpler than those previously employed. Journal articles, newspaper articles, video recordings, monographs, encyclopedia articles, and U.S. Government Documents are included.

Scope: International.

Locating Material: The reader is advised to scan the alphabetical list of subject headings in the Table of Contents to determine where to find the subjects of interest, then search under those subjects. There are numerous cross references to lead to alternate terms.

Abstracts: Not applicable. Author, title, and full bibliographic information is given. The number of references and footnotes and a bioethics accession number (which is for internal use by the bibliography staff) and a list of descriptors are given.

Indexes: Author and title indexes are provided in each volume. The reference is to page number.

Other Material: A detailed discussion of the bibliography and a description of its use is found in each volume. A list of topics and subtopics is also given.

Periodicals Scanned: A list which gives the International Standard Serials Number (ISSN) when available is provided in each volume.

Database: Citations are available through BIOETHICSLINE from 1973. The update is three times a year and keywords are provided.

BIBLIOGRAPHY OF REPRODUCTION, 1963-
Reproduction Research Information Service, Ltd. Monthly.

Arrangement: Arranged by subject. Within each subject section the references are arranged according to the animal species concerned, with the higher order of evolution first (man and unspecified species are first; fishes are last). Within these groups one finds 1) books, 2) book chapters, and 3) journal articles arranged alphabetically by journal.

Coverage: Literature from biological, agricultural, medical, and veterinary science sources is examined in order to index material on vertebrate reproduction. Such aspects as environment, nutrition, nervous system, male and female reproductive tract, androgens, progestrogens, fertility and infertility and artificial and controlled breeding are included.

Scope: International.

Locating Material: Each entry is given a consecutive serial number (these began with Volume 1, no. 1 with 30000) which appear in only one subject section (the citation may appear in more than one). The Contents page or Guide to Subject Sections will lead the user to the proper section to scan. The serial numbers are referred to in the indexes.

Abstracts: Not applicable. Entries give all authors' names and the senior author's address. The titles are in English. If the language of the article is other than English, an abbreviation indicating the name of the language is given and summaries in still other languages are noted. Papers with an asterisk indicate that it has from 40 to 100 references.

Indexes: An author and animal index is found in each issue. A semiannual subject index is also available. A search through this index can be more efficiently made using a process called intersection. This is explained in the preliminary pages of the index.

Other Material: A list of future meetings is included from time to time. Descriptions of how to use the journal and the indexes are provided. Abbreviation codes for languages are found. Case reports are sometimes included as well as abbreviations used in the authors' addresses. A list of bibliographies available is provided.

Periodicals Scanned: No comprehensive list is provided, but the sources for material for each issue are given at the beginning of the subject section.

BIOLOGICAL ABSTRACTS, 1926-
BioSciences Information Service of Biological Abstracts. Semimonthly.

Arrangement: Arranged by subject. A comprehensive subject classification outline is provided in the first issue of each volume as well as a list of section headings in each issue. The subject classification outline is also available separately.

Coverage: An extremely comprehensive coverage of the field of biological science. The searcher should consult the subject classification outline or subject guide. There are presently over 600 subject categories. Reports, proceedings, symposia, books, etc. are included.

Scope: International.

Locating Material: The abstracts are numbered consecutively through each volume. These numbers are referred to in the indexes.

Abstracts: The abstracts are in English. The titles are in English. The language of the article and summaries in other languages are indicated. Up to ten authors are named and the senior author's affiliation is given. Illustrative material is noted. Abstracts generally give a good indication of the content of an article and range from very concise to quite detailed. Frequent "see" and "see also" references guide the user to alternative terms.

Indexes: Five indexes are provided:

AUTHOR. Lists alphabetically up to ten personal authors and corporate authors.

BIOSYSTEMATIC. Lists references according to taxonomic categories and within each category by major concept. Each organism mentioned in an article is included--it need not be the primary subject. Asterisks indicate new taxa described or identified.

GENERIC. Lists references by Genus Species names. The major concept emphasis is shown in abbreviated form.

CONCEPT (formerly CROSS or Computer Rearrangement of Subject Specialties). Enables the searcher to discover subject concepts covered in an article other than that of major emphasis. References are found under every subject heading that indicates the content of an article and the searcher can discover relationships among articles. Ten columns are printed across the page headed by the last digit of the reference number (abstract number), i.e., 0, 1, 2, 3, etc., through 9. All reference numbers whose last digit number is zero are found under that column, all ending in one are found under the number 1, etc. One should select the proper subject heading, make note of the reference numbers and match these with numbers found under other headings selected.

SUBJECT (B.A.S.I.C.--Biological Abstracts Subjects in Context). Consists of every significant term (keyword) in the title plus significant terms selected from the article to expand or clarify the title. The index term is found in the center of each column. Text to the left and right indicates subject context. The title is repeated in the index (in permuted form) until all key terms have been positioned alphabetically. Modifying terms may suggest additional search

116 / Abstracts and Indexes

terms to the user. Fragments of words may appear at either end of the line because each line has a limited number of letters and spaces. The number at the end of each line is the entry number for the abstract.

These indexes appear in each issue and cumulate semiannually and at the end of each volume.

Other Material: Biological Abstracts/RRM (Reports, Reviews, Meetings) began with Volume 18, no. 1 (1980). Entries are arranged by section headings and the five indexes aid the searcher in locating material. Complete bibliographic citations are given. Books, book chapters, bibliographies, reports, reviews, and meetings are found here. BioResearch Index was published from 1965-1979 and provided access to research papers in addition to those found in Biological Abstracts. Bibliographic citations, four indexes, a list of journals and a list of abbreviations were given. Symposia, meetings, congresses, reviews, semipopular and trade journals, selected government reports, book editorials, etc. were found. A list of journals covered was provided in each issue. One found the journal source listed once and set off by dotted lines. The volume number and date were given here. Each entry was numbered and arranged sequentially under the journal title. Authors, title (and added keywords) and pagination were given. Entry numbers were referred to in all the indexes. BIOSIS exerts great effort in explaining the use of the abstracts, the indexes, and the other services. These explanations are found throughout the issues. A "Guide to the Preparation of Abstracts" is provided, as is a list of acceptable abstract abbreviations and a key to representation of scientific names of organisms. A "Guide to the Vocabulary of the Biological Literature" is also available separately.

Periodicals Scanned: Serial Sources for BIOSIS Data Base is a list of serial publications which contribute to BIOSIS products. This includes current, ceased, or changed serials and a list of publishers and addresses. Full title, CODEN, frequency and a number which leads one to the publisher's address are given.

Database: Citations are available through BIOSIS PREVIEWS from 1969. Both Biological Abstracts and Biological Abstracts/ RRM (formerly: BioResearch Index) are included. Updates are monthly and abstracts and keywords are provided. Also available is BIOCODES which is a dictionary file for the taxonomic and category (or concept) codes used in BIOSIS PREVIEWS.

BIOLOGICAL MEMBRANE ABSTRACTS see BIOCHEMISTRY ABSTRACTS

CARBOHYDRATE CHEMISTRY AND METABOLISM ABSTRACTS, 1973-1975. Information Retrieval, Ltd. Ceased. Formerly: Carbohydrate Metabolism Abstracts, 1973-1973.

Arrangement: Arranged by subject. Broad areas are subdivided into more narrow topics.

Coverage: The broad subject areas are: structural and functional aspects, metabolic pathways and regulation, transport, enzymes and their regulation, studies on organisms, diseases and metabolic defects (see Contents for a complete list).

Scope: International.

Locating Material: Entries are in numerical sequence through the volume.

Abstracts: Abstracts are in English. Titles are in English and in the language of the article for European material (except that in Cyrillic alphabets). The language of an article and of summaries in translation are indicated. All authors' names are given and the address of the senior author is provided. The author's summary is used when possible. Abstracts are usually informative and range in length from 150 to 300 words. Indicative abstracts are provided for extremely long review papers.

Indexes: An author index is in each issue which cumulates annually. An annual volume subject index in which abstracts are indexed by up to six entries is also provided.

Other Material: Book Notices and Notification of Proceedings are in each issue. Keys to abbreviations found in the abstracts and for languages are provided in each issue.

COMMERCIAL FISHERIES ABSTRACTS see MARINE FISHERIES ABSTRACTS

CURRENT ADVANCES IN GENETICS, 1976-
Pergamon Press, Ltd. Monthly.

Arrangement: Arranged by subject.

Coverage: In the preliminary pages is a section called "Aims and Scopes" which states that this journal "...provides a monthly current awareness service for geneticists, plant and animal breeders, human geneticists, molecular biologists and medical scientists trying to keep abreast of the ever-increasing literature, both journal and books, in the whole field of genetics." Among the topics listed are: genetic code, DNA replication, recombination, protein synthesis, cell division and the cell cycle, biochemical, developmental and behavioural genetics, animal and plant breeding, animal and plant viruses, medical cytogenetics, and cancer research. There are 54 sections altogether.

Scope: International.

Locating Material: Entries are numbered consecutively through the volume.

Abstracts: Not applicable. Full bibliographic detail is provided with extensive cross-referencing by full title. Reviews of new books are included.

Indexes: An author index is in each issue.

Other Material: "Notification of New Books" is found in each issue. Very brief explanatory material is found.

Periodicals Scanned: A list is found in each issue.

CURRENT ADVANCES IN PLANT SCIENCE, 1972-
Pergamon Press, Ltd. Monthly.

Arrangement: Arranged by forty-four broad subjects most of which are divided into smaller areas.

Coverage: The more general aspects of plant science are treated with subtopics relevant to each subject heading. These headings include photosynthesis, respiration, cell organization and division, membrane structure and properties, vegetative and reproductive development of seed plants, germination, taxonomy, genetics, plant breeding, agronomy, horticulture, tree growth, ecology, and conservation and pollution (see Contents for a complete list).

Scope: International.

Locating Material: Entries are numbered consecutively from the beginning of the publication. These numbers are referred to in the indexes.

Abstracts: Not applicable. All authors are given and the address of the senior author is provided. The titles are in English. A list of relevant titles is given at the end of each subject section.

Indexes: An annual author index is provided.

Other Material: Short articles or commentaries are included in many issues. A list of new books (Notification of New Books) is found in each issue.

Periodicals Scanned: A list is given in each issue of titles searched or that issue.

DEEPSEA RESEARCH, 1953-
Pergamon Press. Monthly. In 1979 divided into Part A, Oceanographic Research Papers, and Part B, Oceanographic Literature Review (which was previously, the Oceanographic Abstract and Bibliography Section). Formerly: Deepsea Research and Oceanographic Abstracts.

Arrangement: Part A contains articles reporting results of research, solution of instruments problems, a new laboratory method, etc. An abstract precedes each article. Paging for each part is separate and continuous through each volume. Part B, the Literature Review, is arranged by broad subjects which are subdivided into more narrow topics. Each broad subject is assigned a letter and each topic is given a number (before 1979 this section was paged separately and continuously and began each volume with "A1"). It is divided into Pt. 1 and Pt. 2. The date of receipt of a citation determines its inclusion in either Pt. 1 or Pt. 2. Each part is divided into two sections: titles and bibliographic references (arranged by subject) and abstracts (arranged alphabetically by author).

Coverage: Physical oceanography including currents, waves, tides and sealevel, marine meteorology, submarine geology and geophisics, chemical and biological oceanography are included.

Scope: International.

Locating Material: Reviews are given numbers. The first two digits indicate the year and the others are the citation number; the arrangement is sequential through each volume. These numbers are referred to in the indexes.

Abstracts: Not really applicable in that the citations are actually short reviews. Nevertheless, concise information is given about the contents of the articles. Occasionally no review appears.

Indexes: Quarterly author and subject indexes are provided. The terms employed are those assigned to each document and are drawn from the title or text and are rotated to generate the index entries.

Other Material: A detailed description of the subject index is provided each time it appears, in which hierarchies and cross references are explained and a search strategy is suggested.

Periodicals Scanned: No list is provided. The citations are those received by the editors within a given time period for inclusion in a certain issue.

ENTOMOLOGY ABSTRACTS, 1969-
Cambridge Scientific Abstracts. Monthly.

Arrangement: Arranged by broad subjects subdivided into smaller topics.

Coverage: All aspects of the subject are covered. Abstracts of published proceedings where the full paper is published are included. Specific subjects include systematics, technique, morphology, physiology, anatomy, histology, and biochemistry, reproduction and development, behavior, biology and ecology, genetics, geography and present-day faunas; fossil forms and faunas are listed as broad topics in the Table of Contents.

Scope: International.

Locating Material: Entries are numbered consecutively.

Abstracts: Titles and abstracts are in English. Titles of papers in the European languages are also given in the original language; for other languages only the English translation is given. The language of the article is indicated, and if there are summaries available in translation, this information is also provided. The address of the senior author is noted; if no address is given in the primary source the editorial address of the latter is given. Articles appear in the section to which they are most relevant and are cross-referenced to other sections. The author's abstract is used when it is editorially acceptable although it may be altered slightly. In all other cases, abstractors prepare a summary based as far as possible on an abstract or summary in the source journal.

Indexes: Each issue has author and subject indexes which cumulate annually.

Other Material: A list of abbreviations used in the text and of languages appears in each issue. Explanatory material is found in the first issue of the volume.

Periodicals Scanned: A list is supplied on request. A separate list covering Cambridge Scientific Abstracts life science abstract journals is available for sale.

Database: Citations are available through IRL LIFE SCIENCE COLLECTION from 1978. There are monthly updates and keywords and abstracts are available.

EXCERPTA BOTANICA, 1959-
Gustav Fischer Verlag. Published in two sections: A-Taxonomia et Chorologica; B-Sociologica. Section A has 1-2 volumes per year (seven numbers per volume) and Section B has one volume per year (four numbers per volume).

Arrangement: Section A is more or less by subject, being divided into several parts (which are also designated by letters as the two major sections are). Pt. A; Taxonomia et Phylogenia with subsections for narrower subjects; Pt. B: Chorologica with subsections for geographical areas; Pt. C: Paleobotanica; Pt. D: Varia; and Pt. E: Recensiones are the sections encountered. Section B varies and may be by author, by subject, or chronological.

Coverage: A broad range of botanical literature including national and subject bibliographies, periodical articles, and book reviews is found. The user should refer to the contents of the issue of all sections for the subject matter covered.

Scope: International.

Locating Material: No numbering system is used. One must depend on the Contents or Indexes or go to the proper subject section, geographical area, period of time, etc.

Abstracts: In Section A, one may expect to find titles and abstracts in English, French, or German. Senior and joint authors' names are given, illustrative material is indicated, language of the article and summaries in other languages are sometimes noted. The abstractor's name and the city he is from are given. Abstracts are usually brief and occasionally none will be found. No abstracts appear in Section B. One will find subject bibliographies and bibliographies of geographical areas. Entries are usually in the language of the publication.

Indexes: Section A has an annual author index, index to scientific plant names and a geographical index. Section B has an author index for each bibliography in each issue and a subject index where necessary.

Other Material: A list of abstractors with their addresses and the countries for which they are reporting is provided.

GENETICS ABSTRACTS, 1966-
Cambridge Scientific Abstracts. Monthly.

Arrangement: Arranged by broad subject areas subdivided into smaller topics. In the subdivisions, abstracts are in order according to species.

Coverage: The Preface states: "All aspects of molecular, viral, bacterial, fungal, algal, plant, animal and human genetics...." Subjects include chromosomes, extrachromosonal genetics, cell division; chemical mutagenesis; radiation genetics; developmental, evolutionary, ecological, behavioral, theoretical, and population genetics, among others.

Scope: International.

Locating Material: Entries are numbered consecutively through the volume. First is the sequential number then a code letter identifying the journal and a number for the volume.

Abstracts: Abstracts are in English. Titles are in English and in the language of the article for material in European languages except those using Cyrillic alphabets. For non-European languages, titles are in English only. The language of the article and summaries in translation are noted. Up to ten authors are named and the address of the institute of the first author is given. Cross references are provided to lead the searcher to relevant subheadings.

Indexes: Author and subject indexes are in each issue. These cumulate annually.

Other Material: Lists of abbreviations appear in each issue. Explanatory material is found in the first issue of the volume.

Periodicals Scanned: A list is available on request. A separate list of Cambridge Scientific Abstracts life science abstract journals is available for sale.

Database: Citations are available through IRL LIFE SCIENCES COLLECTION from 1978. There are monthly updates and abstracts and keywords are provided.

HELMINTHOLOGICAL ABSTRACTS, 1932-
Commonwealth Agricultural Bureaux, prepared by Common-

wealth Institute of Parasitology. Quarterly 1932-1973. From 1973, Ser. A., Monthly; Ser. B. Quarterly. Volume 1-3, no. 4 published as Supplement to Journal of Helminthology. With Volume 39, divided into Ser. A, Animal and Human Helminthology and Ser. B, Plant Nematology.

Arrangement: Arranged by subject.

Coverage: Ser. A covers animal hosts, morphology, taxonomy, immunity, ecology, geographical distribution, etc. and Ser. B, Plant hosts, Arthropod hosts, marine Nematodes, morphology, biology, ecology, geographical distribution, etc. (See Table of Contents on the inside front cover of both series.) Books and nonperiodical material, conferences, congress proceedings, and reports are carried at the end of each issue.

Scope: International.

Locating Material: The entries are numbered consecutively through each volume. Indexes refer to these numbers.

Abstracts: Titles are in English and usually in the language of the article. If summaries in languages other than that of the article are available, this is noted. Occasionally the title will be given only in English, but the language of the article will be specified. Abstracts, which are usually detailed, are in English. Sometimes no abstract appears. The language of the article and of summaries is indicated by abbreviations and the journal citation is given in full. The number of references is noted and the affiliation of the senior author is given if available.

Indexes: Author and subject indexes are in each issue and cumulate annually.

Other Material: A review article is found in some issues. A list of publication in Animal Helminthology and Plant Nematology is provided in the appropriate series from time to time. An explanation of abbreviations and language symbols is found in the preliminary pages of each issue.

Periodicals Scanned: The latest list at this writing is found in Volume 50, no. 1. This list can be obtained separately from the Commonwealth Institute of Parasitology.

Database: Citations are available through CAB ABSTRACTS from 1973. These are updated monthly and contain abstracts and keywords. The other abstract journals of the Commonwealth Agricultural Bureaux are also included.

124 / Abstracts and Indexes

IMMUNOLOGY ABSTRACTS, 1976-
Cambridge Scientific Abstracts. Monthly.

Arrangement: Arranged by broad subjects subdivided into narrower topics.

Coverage: The broad topics include molecular immunology and methodology, immune responses, immunomodulation, immunity to infection, tumor immunology, histocompatibility and transplantation, and immune disorders. Books, proceedings, and forthcoming meetings are included.

Scope: International.

Locating Material: Entries are numbered consecutively through each volume. There are three parts to these numbers: first a number for the abstract, then a code letter identifying the journal, and finally the volume number.

Abstracts: Titles are in English and in the language of the article. The language and summaries in other languages are indicated.

Indexes: There are author and subject indexes in each issue which cumulate annually.

Other Material: Descriptive information and material explaining how to use the journal is provided. A list of abbreviations is found.

Periodicals Scanned: A list of periodicals searched is available on request. A separate list covering all the Cambridge Scientific Abstracts life science abstract journals is available for sale.

Database: Citations are available from 1978 through IRL LIFE SCIENCES COLLECTION and are updated monthly. Keywords and abstracts are provided.

INDEX TO AMERICAN BOTANICAL LITERATURE in: TORREY BOTANICAL CLUB BULLETIN, Vol. 13- (1886-). Torrey Botanical Club. Bimonthly. Index to American Botanical Literature, 1886-1966 and first supplement, 1967-1976 available separately from G.K. Hall. Also printed on cards. Annual, 1959-1968 on cards.

Arrangement: Arranged by subject, then alphabetically by author under subjects.

Coverage: Taxonomy, phylogeny, and floristics of the algae, fungi, bryophytes, pteridophytes, and spermatophytes; ecology and plant geography, genetics, morphology, anatomy, cytology, and genetics of these groups; paleobotany, plant ecology, and general botany (which includes biography and nomenclature) are covered.

Scope: Material from North, South, and Central America and the West Indies is found.

Locating Material: No numbering system is used. Under each subject section the authors are listed alphabetically.

Abstracts: Not applicable. Entries include all authors' names, but cross-references from joint authors are not provided. The title is in the language in which it is published. The journal citation is in English.

Indexes: Index to American Botanical Literature, 1886-1966 and first supplement, 1967-1976 are available separately from G.K. Hall. Annual, 1959-1968 on cards. A supplement in book form is planned every ten years.

INTERNATIONAL ABSTRACTS OF BIOLOGICAL SCIENCES, 1954- Pergamon Press. Monthly. Four volumes per year. Supersedes British Abstracts, Ser. A, Sec. 3. Title varies: British Abstracts of Medical Sciences, Vols. 1-3, 1954-1956.

Arrangement: Arranged by broad subject.

Coverage: "Covers the more important papers in biochemistry, physiology, pharmacology, microbiology, immunology, oncology, cell biology, genetics, anatomy, animal behaviour, and experimental zoology." (Aims and Scope.)

Scope: International.

Locating Material: Items are numbered consecutively through each volume.

Abstracts: Abstracts range up to 100 words in length. In some cases, if the paper is a preliminary communication or if the subject is borderline, no abstract is given. Abbreviations of journals cited in the abstracts are based on those used in World List of Scientific Periodicals. Full bibliographic information and the first author's affiliation is given.

Indexes: Each issue has an author index.

Other Material: The Table of Contents is a list of the subject headings used with cross-references to other relevant sections. The first issue of each volume has a list of abbreviations used. In each issue there is a section which is a list, following the abstracting section, of titles appearing in issues of society proceedings, symposia, and other publications appearing at intervals which make them fall outside the usual definition of a journal.

Periodicals Scanned: A list appears in the first issue of each volume. It is divided into two sections; the journals in List A are covered by abstracts, titles, or expanded titles, and those in List B are covered by titles or expanded titles.

KEYWORD INDEX OF WILDLIFE RESEARCH, 1974-
Swiss Wildlife Information Service. Annual.

Arrangement: The index is divided into two parts. Part 1 is arranged by keyword (with an author and title index) and Part 2 is a thesaurus and a number of lists.

Coverage: Behavior, conservation, diseases, ecology, food, morphology, national parks, natural resources, parasitology, pollution, population, wildlife management, and zoogeography are among the subjects covered.

Scope: International.

Locating Material: The keyword index should be used for finding the material the user is interested in; then the author index should be consulted for full bibliographic information.

Abstracts: Not applicable. The entries in the keyword index use only the common name of a species. A list of species should be used if there are two names or if only the scientific name is known (Part 2, list D). Full bibliographic information is found in the author index.

Indexes: Author and title indexes are found in each issue.

Other Material: There is explanatory material at the beginning of the indexes and lists which gives helpful information. The thesaurus gives synonyms and references to related entries and the species list gives English, German, French, and Latin names of species included in the keyword index.

Periodicals Scanned: A list is provided in each issue.

MARINE FISHERIES ABSTRACTS, 1948-1974.
U.S. Department of Commerce, National Oceanic and Atmospheric Administration, National Marine Fisheries Service (U.S. Government Printing Office). Formerly: Commercial Fisheries Abstracts. Ceased.

Arrangement: Arranged by classification (code) numbers. Entries are printed in blocks (from side to side on the page rather than from top to bottom) which are the size of standard 3 x 5 inch cards.

Coverage: Biological aspects of fishery science and technological studies dealing with aquatic resource supply, harvesting, processing, utilization, and distribution. Annual reports, patents and proceedings are included.

Scope: International. Includes trade, engineering and scientific journals.

Locating Material: Searcher should refer to the author or subject index to find a code number and a page number where the reference is located. A note on the inside back cover states that: "Code numbers used to identify the subject of abstracts are translated in Fishery Leaflet No. 232, Fishery Technological Abstract Card System."

Abstracts: Abstracts are in English. The titles are in the language of the article and in English. Summaries in other languages are indicated. The senior author's affiliation is given. Abstracts are usually quite detailed. If the abstract is from another source, the source is given. Periodical titles are given in the language of the periodical and in English. Subject code numbers and Marine Fisheries Abstract volume, issue, and page numbers are given for each entry. The number of references and illustrative material are indicated.

Indexes: Author and subject indexes are in each issue.

Other Material: A brief Foreword and order information are on the inside front and back covers of each issue.

Periodicals Scanned: The list appears in Marine Fisheries Abstracts about once a year.

MICROBIOLOGY ABSTRACTS, 1966-
> Cambridge Scientific Abstracts. Monthly. Published in three sections: A, Industrial and Applied Microbiology (title of section varies: Industrial Microbiology); B, Bacteriology (title of section varies: General Microbiology and Bacteriology); (both sections supersede and continue the volume numbering of Industrial Microbiology Abstracts); C, Algology, Mycology, and Protozoology (published from 1972). The description below covers all three sections.

Arrangement: Arranged by broad subjects which are further subdivided into narrower topics. A full explanation of the arrangement of the sections is found in the Preface to each issue.

Coverage: Section A Contents lists such topics as products of microorganisms, fermentation and related processes, microbial degradation, food microbiology, microbial toxins, plant diseases, forestry, soil microorganisms, mineral microbiology, antimicrobial agents, vaccines, and environmental pollution. Section B includes methodology, cell structure and function, genetics and evolution, antibiotics, aggressins and toxins, immunology, human, animal and invertebrate bacteriology, plant diseases, and ecology and distribution. Section C provides coverage for taxonomy, structure and function; growth, development, and life cycles, ecology and distribution, biochemistry, nutrition, genetics, mycotoxins, immunology and vaccination, parasitism and diseases, and viruses and bacteria of microorganisms. The Contents of each section furnishes a complete list with subheadings under the broader topics.

Scope: International.

Locating Material: In each section the abstracts are numbered consecutively through the volume. The abstract number carries the volume number, a code letter which identifies the abstract journal and the sequential number of the citation.

Abstracts: Titles and abstracts are in English. Titles of papers in other languages are also given in the language of the article. Up to ten authors' names are given and the address of the senior author is provided. Journal titles are abbreviated in accordance with the procedures laid down by the American National Standards Institute. Abstracts run from 150-200 words except where longer ones are needed. The author's abstract is used when feasible. Citations are cross-indexed to all subheadings to which they are relevant.

Indexes: There are author and subject indexes in each issue which cumulate annually.

Other Material: A list of abbreviations is found in each issue of each section. Each has a table of contents which shows the full scope of the subjects covered in the abstracts. Book Notices and Notification of Proceedings also appear in each issue of each section.

Periodicals: A list is available on request.

Database: Citations are available through IRL LIFE SCIENCES COLLECTION from 1978. The updates are monthly and abstracts and keywords are provided.

NUCLEIC ACIDS ABSTRACTS see BIOCHEMISTRY ABSTRACTS

OCEANIC ABSTRACTS, 1964-
Cambridge Scientific Abstracts. Bimonthly. Supersedes: Oceanic Index and Oceanic Citation Journal.

Arrangement: Arranged by subject.

Coverage: The subjects listed in the Contents are: marine biology and biological oceanography; physical and chemical oceanography and meteorology; marine geology, geophysics, and geochemistry; marine pollution; living and nonliving marine resources; and ships and shipping.

Scope: International.

Locating Material: The entries are numbered consecutively through each volume. The first two digits indicate the year and the next five are the sequential number. These begin in each volume with 0001.

Abstracts: The titles are in English and usually in the language of the article. An abbreviation indicating that language and any available summaries is given. The author's address indicates where the work was carried out or where correspondence should be addressed. In the introductory information one finds: "The abstracts, which are up to 200 words long, give the basic information reported in the original material.... Citations only coverage is given to revelant papers in symposia, to individually authored chapters in books, and to notes, letters, and preliminary research communications...."

Indexes: Each issue contains an author and a subject index and from 1980 an organism index (which uses the scientific name). There is an annual cummulative index.

Other Material: There is explanatory information, a list of acronyms, and a list of abbreviations, prefixes and symbols in the preliminary pages of each issue.

Periodicals Scanned: There is a list provided in the 1980 cumulative index.

Database: Citations are available through OCEANIC ABSTRACTS from 1964. The update is every two months and keywords are provided. Abstracts are also provided but only from 1978.

PROTOZOOLOGICAL ABSTRACTS, 1977-
Commonwealth Agricultural Bureaux. Monthly.

Arrangement: Arranged by subject with subdivisions for narrower topics.

Coverage: Hosts, taxonomy and nomenclature, morphology and ultrastructure, evolution and genetics, immunity, serology and resistance, pathology and pathogenicity, biochemistry, physiology and behavior, epidemiology and epizootiology are among the subjects listed in the Table of Contents.

Scope: International.

Locating Material: Entries are numbered consecutively through each volume. These numbers are referred to in the indexes.

Abstracts: Titles and abstracts are in English and if the article is in another language the title is repeated in that language except for Chinese, Japanese, Arabic, etc. Titles of works in the Cyrillic languages are transliterated for books only. The language and summaries in other languages are indicated. Illustrative material and the number of references is noted.

Indexes: Author and subject indexes are found in each issue. These cumulate annually.

Other Material: Occasionally a review article will appear. Beginning with Volume 7, a "Reader's Guide" which describes the journal is provided. A list of abbreviations for languages and "common units and chemical elements, etc." is provided.

Periodicals Scanned: A list appears from time to time. There is usually one with each volume.

Database: Citations are available through CAB ABSTRACTS from 1977. They are updated monthly and abstracts and keywords are provided. The other abstract journals of the Commonwealth Agricultural Bureaux are also included in the database.

REVIEW OF APPLIED ENTOMOLOGY, 1913-
  Commonwealth Agricultural Bureaux, prepared by Commonwealth Institute of Entomology. Ser. A, Agricultural; Ser. B, Medical and Veterinary. Monthly.

Arrangement: Presently arranged by broad subjects with subdivisions for narrower topics. Early volumes had a random arrangement.

Coverage: Covers all aspects of applied entomology, the one series including agricultural and the other medical and veterinary information. In addition to journals, documents (state and national), reports, conferences, congresses symposia, and publications from such bodies as the World Health Organization and others are found. Subjects listed in the Contents for both sections include taxonomy, anatomy, morphology, reproduction and development, physiology and biochemistry, genetics and sterility, ecology and behavior, geographical distribution, etc.

Scope: International.

Locating Material: Entries are numbered consecutively through each volume. Square brackets around a number indicate that the abstract is one of a series with the same bibliographic data as that of the nearest preceding number with no bracket.

Abstracts: Titles and abstracts are in English. The language of the article and summaries in other languages are noted. The senior author's affiliation (complete address), number of references, and illustrative material are noted.

Indexes: Author and subject indexes are provided in each issue.

Other Material: Occasionally there is an editorial which may be a short news notice or a long article. The same editorial is carried in both sections. Brief explanatory material and a list or abbreviations are provided.

Periodicals Scanned: A list appears from time to time, usually in the first issue of a volume.

Database: Citations are available through CAB ABSTRACTS from 1973. Monthly updates are provided as are keywords and abstracts. The other abstract journals of the Commonwealth Agricultural Bureaux are also included.

REVIEW OF PLANT PATHOLOGY, 1922-
Commonwealth Agricultural Bureaux, prepared by Commonwealth Mycological Institute. Monthly. Formerly: Review of Applied Mycology, 1922-1969.

Arrangement: Arranged by subject. Some topics are subdivided for narrower categories.

Coverage: The more important literature on "diseases of plants caused by fungi, bacteria, viruses, mycoplasmas, and nonpathogenic factors" is covered. Included are books and other reports on conferences, congresses, symposia, etc. Taxonomic papers are listed in the Institute's half-yearly Bibliography of Systematic Mycology, and new generic and specific names of fungi are compiled in Index of Fungi. Subjects listed in the Contents include bacteria, fungicides, antibiotics, legislation, soils, viruses, and crops (both general and specifically named crops).

Scope: International.

Locating Material: Entries are numbered consecutively.

Abstracts: Titles and abstracts are in English; titles of papers in European languages, including those using the Cyrillic alphabet are also given in the original language; for other languages only the English translation is given. Cyrillic characters are transliterated according to International Standards. If necessary, the names of pathogens are adjusted to those used in the Review. Summaries in translation are indicated. The number of references for papers providing a useful literature review and illustrative material are indicated. One may find references to earlier issues of the Review or to the Institute's Distribution Maps of Plant Diseases when the record is a new one. The address of the first author is given when known and if the date of publication of a citation is in question, its date of receipt will be given.

Indexes: Author and subject indexes are found in each issue. These cumulate annually.

Other Material: Explanatory material is found in the first issue of the volume. There is a list of abbreviations, a list of addresses and of liaison officers provided.

Database: Citations are available through CAB ABSTRACTS from 1973. There are monthly updates and keywords and abstracts are provided. The other abstract journals of the Commonwealth Agrcultural Bureaux are also included.

SPORT FISHERY ABSTRACTS, 1955-
U.S. Department of the Interior, Fish and Wildlife Service (U.S. Government Printing Office). Quarterly.

Arrangement: Arranged by subject. Broad topics are subdivided into narrower categories.

Coverage: Aquatic plants and their control, culture and propagation of fish, limnology and oceanography, morphology, physiology, genetics and behavior, natural history, parasites and diseases, pollution and toxicology, and research and management. (See Contents for a complete list with subdivisions.)

Scope: International.

Locating Material: Abstracts are numbered consecutively through each volume. These numbers are referred to in the indexes.

Abstracts: Abstracts are in English. Titles are in English and the language of the article. Summaries in other languages are indicated. Affiliation of the senior author is given. If the abstract is from another source or from the author, it is so stated. Generally, the abstracts are detailed and sometimes quite lengthy, but occasionally none appears.

Indexes: Author, geographic, and systematic (taxonomic) indexes are in each issue and in the final issue there are cumulative author, geographic, taxonomic and subject indexes. Cumulative indexes to Vol. 1-5, 6-10, 11-15 are available separately.

Periodicals Scanned: A list is found in the final issue of each volume.

TISSUE CULTURE ABSTRACTS, 1963-
Grand Island Biological Co. (GIBCO Laboratories). Bimonthly.

Arrangement: This work is divided into two parts: Mammalian tissue culture and insect tissue culture (insect tissue culture does not always appear) and within each section by date.

Coverage: All aspects of tissue culture from mammals, insects and plants are covered in depth. Research topics, medical treatment, diseases, and disorders are included.

Scope: International.

Locating Material: Entries are numbered consecutively from Volume 5.

Abstracts: Titles are in English. All authors and their affiliations are provided. The abstracts are usually quite detailed giving a good idea of the content of the article.

Periodicals Scanned: A list is found in each issue.

WILDLIFE ABSTRACTS see WILDLIFE REVIEW

WILDLIFE REVIEW, 1935-
U.S. Department of the Interior, Fish and Wildlife Service (U.S. Government Printing Office). Quarterly.

Arrangement: By subject. Broad areas are subdivided into smaller groups. Entries are alphabetical by author within each subject.

Coverage: General conservation, soils, plants, wildlife, pollution, food habits, mammals (general and specific species), birds, reptiles, etc. are included (see Table of Contents).

Scope: International.

Abstracts: Not usually provided. Sometimes only an author-title citation is given and sometimes an indication of the contents is found. Citation for available abstracts is given if an entry is taken from an abstracting journal. The author's address is provided when known. The title is given in the language of the article and in English. The language is specified and any summaries in other languages are indicated.

Indexes: An author, subject, and geographic index is found in each issue. (See Other Material below.)

Other Material: Wildlife Abstracts is published from time to time as a bibliography and index to Wildlife Reviews. Years covered are 1935-1951, 1952-1955, 1956-1960, 1961-1970. The first section is a bibliography of titles by subject and the second section contains indexes. The subject groups are identified by a three-digit code number. This leads the searcher to the proper section in the bibliography for a given subject. Each entry has the code numbers, an abbreviation for the author, and a two-digit identification number (based on the third letter of the senior author's last name). Each citation also has the Wildlife Review number indicating volume and page where it can be found. Author and subject indexes are provided with each issue which list the citations by code number. The subject index is animal oriented. A "List of Bibliographic Sources" is found at the end of the volume which gives the complete title and country of origin of the publications cited (in abbreviated form) in the Bibliography.

WORLD FISHERIES ABSTRACTS, 1948-1973.
Food and Agricultural Organization of the United Nations.
Ceased. Published in English, French, and Spanish.

Arrangement: Arranged by a code for abstract numbers (see below) which forms a loose, informal subject arrangement. The abstracts are printed on international size (127 x 76 mm) cards and appear in the abstracts uncut. To use in the uncut form, indexes provided in each issue are necessary. The cards may be cut and filed by any of five possible methods described.

Coverage: All aspects of technical literature on fisheries and related industries including canning and freezing methods, fishing boats and equipment, fishing methods, water purification, kinds of fish, etc. are found (see subject index).

Scope: International.

Locating Material: Each abstract is numbered with a six-digit figure as follows: two digits indicating the volume number, a hyphen, a digit denoting the number of the issue, two digits denoting the page in the issue and the last denoting the serial position of the abstract on the page; e.g., 23-3033 refers to Vol. 23, no. 3, page 3, third abstract from the top of the page. If the abstracts are kept in the volume uncut, one must refer to the author or subject index which leads him to the proper abstract number. Universal Decimal Classification (UDC) and U.S. Fish and Wildlife Service (FWS) numbers are provided for each abstract. If the abstracts are cut and filed by either of these systems, a cross-index of abstract numbers to the proper classification number is provided.

Abstracts: Titles and abstracts are in English. Usually they are quite detailed, some providing illustrations. The abstractor's name is given. Occasionally more than one title on a specific subject will be on a card but each entry carries its own number. The FWS and UDC classification number is given, as is the World Fisheries Abstracts volume, issue, page numbers, and year. If the article is in a language other than English, the title is bracketed. References and illustrative material are noted. Occasionally the address of the first author is given and the address of the publisher if the journal did not appear in the "List of Periodicals Searched."

Indexes: Each issue has an author and subject index and a cross index of abstract numbers to authors or titles (in the case of anonymous works), and FWS and UDC code numbers. These have been provided since Vol. 15. Explanatory material precedes the indexes.

Other Material: "Directions for Use" is carried in each issue with a brief description of the five filing systems to be used if the abstracts are cut. Handbook for World Fisheries Abstracts, published separately, gives detailed information.

136 / Abstracts and Indexes

Periodicals Scanned: A list is published separately as "List of Periodicals Searched as at 31 December 1959" and issued as "Supplement" to World Fisheries Abstracts, Vol. 11, no. 1, 1960 (amended from time to time).

ZOOLOGICAL RECORD, 1864-
BioSciences Information Service (BIOSIS) and the Zoological Society of London. Annual in twenty issues.

Arrangement: This work is divided into 27 sections in 20 issues representative of the various zoological groups. Each section covers a phylum or class of animals except the first, which deals with literature relating to more than one branch of zoology, and the final section which lists the new genera and subgenera contained in the other sections. The searcher is provided several avenues of approach. The sections begin with an alphabetical arrangement by author, which gives full bibliographic information followed by subject, geographical, palaeontological, and systematic indexes. Each of these follows the same basic pattern--a series of short indicative sentences arranged under appropriate headings--orders or families, in the systematic, countries or regions in the geographical, etc. Within a heading, the entries are alphabetical. Each indicative sentence refers to the relevant aspects of one paper and is followed by the name of the author, thus enabling the searcher to locate the bibliographic details of any interesting items. In the author section an item number is assigned each senior author. This allows the user to find a specific paper if an author has written more than one, from information in the other four sections.

Coverage: Reproduction, development genetics, ecology, geographical distribution, behavior, morphology, palaeontological distribution, etc. of the various zoological groups are covered. (A detailed list of subject headings is given in the preliminary pages of each section.)

Scope: International.

Locating Material: In the author sections, as the arrangement is alphabetical, one merely searches for the name. "See" references are provided from joint authors to primary author. In the other sections, one is referred to the primary author for complete bibliographic information. A Table of Contents and detailed subject index in the preliminary pages of each section refer the user to a page number. If there is more than one citation by an author, the succeeding entries are numbered consecutively (in brackets after the author's name). This number is given in the subject and systematic sections if the reference is to a citation other than the first. In the author section, abstracts and anonymous works are listed first and also carry numbers for easier reference to the other sections.

Abstracts: Not applicable. Author and title citation only is given. Titles are in the language of the article. If summaries are available in other languages this is noted. Illustrative material is noted.

Indexes: The parts of each of the 27 sections are referred to as indexes. An index to genera is provided where appropriate. The headings which appear in the subject, geographical, palaeontological, and systematic indexes are taken from a controlled vocabulary.

Other Material: A detailed description of how to use this journal is found in the preliminary pages of each section. This includes an explanation of the arrangement of the indexes and of the entries in each part.

Periodicals Scanned: A list appears irregularly and may be found in the following volumes: 43, 46, 49, 58, 64, 67, 71, 73, 107, and 112.

Database: Entries are available through ZOOLOGICAL RECORD (produced cooperatively with BIOSIS) from 1978. At this writing, only 1978 is available, but later years are being added and the update frequency will be every two months.

AGRICULTURAL SCIENCES

ABSTRACTS ON TROPICAL AGRICULTURE, 1975-
　　Royal Tropical Institute. Monthly. Formerly: Tropical
　　Abstracts.

Arrangement: Arranged by subject.

Coverage: Soils, fertilizers, and plant nutrition; crop production;
　　animal husbandry; various crops by name; food science and
　　economics are among the subjects listed in the Table of
　　Contents.

Scope: International.

Locating Material: Entries are numbered consecutively through the
　　volumes. The indexes refer to these numbers.

Abstracts: Titles are in English and in the language of the article.
　　The language and summaries in other languages are noted.
　　The abstracts are in English. Illustrative material and the
　　number of references, if significant, are indicated.

Indexes: Author, subject, affiliation, geographic and plant taxonomic
　　name indexes are in each issue. An annual cumulative subject and plant taxonomic index is provided.

Other Material: Extensive explanatory material is provided. Lists
　　of abbreviations used in the abstracts, including those for
　　countries and languages, are provided. Review articles are
　　occasionally included.

Database: Entries are available through TROPAG from 1975 and
　　are updated quarterly.

AGRICULTURAL ENGINEERING ABSTRACTS, 1976-
　　Commonwealth Agricultural Bureaux. Monthly.

Arrangement: Arranged by broad subjects with subtopics.

Coverage: Mechanical power, land improvement, crop production,
　　protected cultivation, crop harvesting and threshing, handling
　　and transport, crop processing and storage, farm buildings
　　and equipment, and aquaculture are covered. Books, con-

ferences, and reports are included. It is stated that the object is to "provide factual summaries of the world scientific and technical literature on agriculture and horticulture machinery, implements, equipment, and buildings, and allied subjects."

Scope: International.

Locating Material: Entries are numbered consecutively through each volume. These numbers are referred to in the indexes.

Abstracts: Titles and abstracts are in English. If the article is in a language other than English, that language is noted by an abbreviation. The title is usually repeated in the other language and summaries in other languages are also reported. The number of references is given, as is the affiliation of the senior author.

Indexes: There is an author index in each issue with an annual cumulation and an annual subject index.

Other Material: "Notes to Readers," includes a list of journal abbreviations, symbols for language, and a few brief explanatory lines. Information about the Commonwealth Agricultural Bureaux is also included.

Periodicals Scanned: A list is provided at intervals, usually in the first issue of a volume.

Database: Entries are available through CAB ABSTRACTS since 1976. Keywords and abstracts are provided and the update frequency is monthly. Other abstract journals of the Commonwealth Agricultural Bureaux are also included.

AGRICULTURAL INDEX see BIOLOGICAL AND AGRICULTURAL INDEX

AGRICULTURAL LITERATURE OF CZECHSLOVAKIA, 1956-
Institute for Scientific and Technical Information for Agriculture. Quarterly.

Arrangement: Arranged by broad subjects with subdivisions for some.

Coverage: Agricultural economics and engineering, plant production with such subtopics as fertilizing, protection, fruit and vegetable growing as well as kinds of crops (cereals, fodder, pulse and technical crops and ornaments); livestock production which treats cattle, horses, sheep, pigs, poultry, rabbits, and fur-bearing animals, fish, and bees, veterinary medicine, forestry, game management, and life environment protection are covered.

Scope: Czechoslovakian literature.

Locating Material: Entries are numbered consecutively through each volume. These numbers are referred to in the indexes.

Abstracts: The titles are in Czech and in English. The abstracts are in English. Illustrative material and the number of references are indicated and if summaries in languages other than that of the article are provided, this is noted.

Indexes: Annual author and subject indexes (in English) are provided.

Other Material: Surveys of studies on reports for the year are provided.

AGRINDEX, 1975-
AGRIS, Food and Agriculture Organization of the United Nations. Distributed in the United States by Unipub. Monthly.

Arrangement: Arranged by broad subject categories, then by narrower topics within a category, then alphabetically by author.

Coverage: Agriculture, economics, rural sociology, plant and animal protection, forestry, aquatic science, natural resources, food science and human nutrition, home economics, pollution, and machinery are included.

Scope: International.

Locating Material: Each entry is given a "Reference Number." These numbers are sequential from volume one and are referred to in the indexes. "See also" references at the beginning of each subject section lead the reader to related material.

Abstracts: Not applicable. All authors' names are given as well as corporate authors where appropriate. The senior author's affiliation is provided. Titles are in English and in the language of the article with an abbreviation indicating what that language is. Information on availability is given when possible (if indicated).

Indexes: A personal author, corporate author, report and patent number, and commodities index are found in each issue.

Other Material: Explanatory information with examples is provided to facilitate the use of this index. There is also a list of abbreviations. Information on obtaining original articles is given.

Database: Citations are available through AGRIS from 1977. The update frequency is monthly. Some keywords are provided and some abstracts, but only on the tape version.

AMERICAN BIBLIOGRAPHY OF AGRICULTURAL ECONOMICS, 1971-1974.
American Agricultural Economics Association, prepared by its Documentation Center. Ceased.

Arrangement: Nine broad subject areas (within the field of agricultural economics) are used, then the entries are arranged alphabetically by author.

Coverage: All aspects of agricultural economics such as marketing, policies and programs, products (demand, supply, price), food and consumer economics, foreign development, production, economics, and farm management, regional and human development, resource economics, and general economics are found.

Scope: Restricted to literature published by U.S. and Canadian agricultural economists (may be in teaching, extension, research, industry or elsewhere), and does not attempt to compete with World Agricultural Economics and Rural Sociology, but complements it. Includes all types of serial publications, mimeographed reports, statistical reports and papers from conferences and symposia. Contains only literature authored or co-authored by an agricultural economist except for some literature on subjects highly relevant to them.

Locating Material: Each citation carries a sequential number followed by a dash and a two-digit number which represents the year of issue, e.g., 81-71 is the eighty-first citation in 1971. Entries in this index refer to citation number only.

Abstracts: Titles and abstracts are in English. If an article is in another language the title is given in English and in the original language. Abstracts are usually medium length and occasionally none will appear. Complete bibliographical citation, National Agricultural Library's call number, and appropriate subject headings (descriptors) of the citation are provided.

Indexes: Author and subject indexes are in each issue. In the subject index, each citation appears under all the descriptors it is assigned.

Other Material: A description of the objectives, scope, format, etc, of the publication appears in each issue.

ANIMAL BREEDING ABSTRACTS, 1933-
 Commonwealth Agricultural Bureaux, prepared by Commonwealth Bureau of Animal Breeding and Genetics. Quarterly 1916-1972, monthly from 1973.

Arrangement: Arranged by broad subject areas subdivided into smaller topics. In each subsection the arrangement is alphabetical by author.

Coverage: Livestock with subsections on breeds, production genetics, etc., laboratory and other mammals, poultry and other birds, general and theoretical genetics, general reproduction, and fish and invertebrates in aquaculture are given, (see Table of Contents). Book reviews, proceedings, papers presented at meetings, and annual reports are included. (See Other Material, below.)

Scope: International.

Locating Material: Entries are numbered consecutively through each volume. The indexes refer to these numbers.

Abstracts: Titles are in the language of the original article and in English. Abstracts are in English. If the article is in another language, the title is bracketed and the language of the article is indicated. Summaries in other languages are noted. If the abstract is taken from another source, a citation to the source is given. Occasionally no abstract will be provided. In most cases the address of the first author is given.

Indexes: Author and subject indexes are in each issue. These cumulate annually. The subject index is divided into three parts: livestock (except poultry), poultry and other birds, other animals, and general subjects.

Other Material: Review articles are found in most issues and a list of abbreviations used in the abstracts is given in each issue. Book reviews and notices, list of books received, proceedings and papers presented at meetings, annual reports and a section called "News and Notes" are carried at the end of each issue. These entries continue the consecutive numbering of the abstracts. The Commonwealth Agricultural Bureaux offer a variety of services in connection with their publications. These include magnetic tapes of a consolidated database (with samples and prices available on request) search services, annotated bibliographies, and a document delivery service.

Periodicals Scanned: A journal list is published in the first issue of even numbered volumes and in Volume 49 (1981). It is missing in Volume 50 (1982).

Database: Citations are available through CAB ABSTRACTS from 1973. These are updated monthly and contain abstracts and keywords. The other abstract journals of the Commonwealth Agricultural Bureaux are also included.

ARID LANDS DEVELOPMENT ABSTRACTS, 1980-
Commonwealth Agricultural Bureaux. Monthly.

Arrangement: Arranged by broad subjects with subtopics.

Coverage: General geography, earth sciences, biological sciences, agriculture, natural resources, energy resource development, and human systems are the broad subjects listed in the Contents.

Scope: International.

Locating Material: Entries are numbered consecutively through each volume. These numbers are referred to in the indexes.

Abstracts: The abstracts are in English. The titles are in English and usually in the language of the article. The language and summaries in other languages are indicated. The number of references is noted if this is significant.

Indexes: Author and subject indexes are found in each issue.

Other Material: A list of abbreviations for languages and for journals is provided with brief explanatory material. A description of Commonwealth Agricultural Bureaux services and products and a list of addresses and liaison officers are provided.

Database: Citations are available through CAB ABSTRACTS from 1980 and are updated monthly. Abstracts and keywords are provided. The other abstract journals of the Commonwealth Agricultural Bureaux are also included.

BIBLIOGRAPHY OF AGRICULTURE, 1942-
Oryx Press. Monthly. Formerly issued by the U.S. National Agricultural Library, and from July 1942-June 1962 under its earlier name, U.S. Department of Agriculture Library and published by the U.S. Government Printing Office, 1942-1969.

Arrangement: Arranged by subject. A list of subject headings is provided in the preliminary pages of each issue.

Coverage: Agriculture in general and allied sciences with such specific fields as animal science, food, human nutrition and home economics, forestry, insect pests and controls, weeds and herbicides, plant physiology and biochemistry, water resources, etc. are covered. A list of subjects is provided in the preliminary pages of each issue. Journal articles, pamphlets, government documents, proceedings, etc. are included. The Bibliography provides access to documents of U.S. Federal Agencies, the FAO reports, state agricultural experiment stations, and state extension services.

Scope: International.

Locating Material: Six-digit ID numbers are provided each entry in the main entry section and used in all the indexes as the referral point to the complete citation. These numbers follow sequentially from 000001 through the year; they begin anew with number one in each issue.

Abstracts: Not applicable. Author and title citations are in English but the language of the article is specified.

Indexes: Monthly subject, geographic, corporate and personal author indexes are provided. Subject and author indexes cumulate annually. The subject index is arranged alphabetically under subject term headings using the vocabulary from the main entry. Titles are sometimes necessarily cut short. The subject word in a title is in italics and a secondary word, which may be before or after the primary word, will be in boldface type. Under a main term, the alphabetical arrangement is by the boldface terms. The identification number is last and leads the searcher to the main entry in the monthly issues.

Other Material: An introductory guide to the arrangement and use of the main entry section and the indexes is provided in the preliminary pages of each issue. An explanation of the annual cumulative indexes is found in the cumulation. A list of language abbreviations is provided and information on the availability of references is given.

Periodicals Scanned: A list is published separately as Agricultural Journal Titles and Abbreviations. There is also a list of journal title abbreviations cited in each issue.

Database: Citations are available through AGRICOLA from 1970. There are monthly updates and abstracts are provided but not keywords.

BIOLOGICAL AND AGRICULTURAL INDEX, 1916-
    H.W. Wilson Company. Monthly (except August) with quarterly and annual cumulations. Formerly: The Agricultural Index (1916-1964). (Material from Biological and Agricultural Index is reproduced with permission of the H.W. Wilson Company.)

Arrangement: Arranged alphabetically by subject, then by title. Book reviews are found together at the end of each issue and the cumulative volume. Entries here are alphabetical by author.

Coverage: An extremely wide range of subject material pertaining to agricultural biology including chemicals, economics, animal husbandry, botany, entomology, genetics, soil science, engineering, agricultural research, environmental science, ecology, food science, marine biology, forestry, veterinary medicine, horticulture, plant pathology, zoology, etc. is covered.

Scope: English language publications.

Locating Material: No numbering system is used. The searcher must use the proper subject term.

Abstracts: Not applicable. Entries provide complete title, author and joint author but only the senior author if there are more than two. Complete bibliographic information is noted and illustrative material is indicated. Frequent "see" and "see also" references guide the user to alternative terms.

Other Material: A list of abbreviations of periodicals indexed, abbreviations used in the citations and an explanation of the entries are found in each issue and in the annual cumulations.

Periodicals Scanned: A list is found in each annual cumulation and in most issues, but occasionally is left out. In cumulations, price, publisher and frequency are given.

COTTON AND TROPICAL FIBRES ABSTRACTS, 1976-
    Commonwealth Agricultural Bureaux. Monthly.

Arrangement: Arranged by subject.

Coverage: Subjects listed include Gossypium Spp. - cotton, Agave Spp. - sisal, corchorus Spp. - jute, hisbiscus Spp. - kenaf and roselle, and other fibres and fibres in general. Reports, conferences, and books are included.

Scope: International.

Locating Material: Entries are numbered consecutively through each volume. The indexes refer to these numbers.

Abstracts: Abstracts are in English. Titles are in English and usually in the language of the article. The language, and summaries in other languages, are indicated. The number of references is noted if this is significant. "See also" references lead the user to related material.

Indexes: Author and subject indexes are found in each issue.

Other Material: A list of abbreviations for languages and for journals is provided with brief explanatory material. A description of Commonwealth Agricultural Bureaux services and products, a list of addresses and liaison officers are provided.

Database: Only the main abstract journals of the Commonwealth Agricultural Bureaux are listed as being in the CAB ABSTRACTS database. However entries in the more specialized journals such as Cotton and Tropical Fibres Abstracts are derived through the same database, so they can also be found. The earliest entry for any of these is 1973; the update is monthly and abstracts and keywords are provided.

CROP PHYSIOLOGY ABSTRACTS, 1975-
Commonwealth Agricultural Bureaux. Monthly.

Arrangement: Arranged by subject.

Coverage: Germination, growth and senescence, reproductive development, tropism and nastic movement, stomatial movement, photosyntheses and respiration, translocation and accumulation, nutrition, nitrogen fixation, enzymes, growth regulators and metabolic inhibitors, membranes, metabolism, temperature and water relations, salinity, pollution, and radiobiology are among the topics listed in the Contents. Reports, conferences, and books are included.

Scope: International.

Locating Material: Entries are numbered consecutively through each volume. These numbers are referred to in the indexes.

Abstracts: Abstracts are in English. The titles are in English and usually in the language of the article. The language and summaries in other languages are indicated. The number of references is noted. The senior author's affiliation is provided.

148 / Abstracts and Indexes

Other Material: A list of abbreviations for languages and for journals is provided with brief explanatory material. A description of Commonwealth Agricultural Bureaux services and products and list of addresses and liaison officers are provided.

Database: Only the main abstract journals of the Commonwealth Agricultural Bureaux are listed as being in the CAB ABSTRACTS database. However, entries in the more specialized journals such as Crop Physiology Abstracts are derived through the same database, so they can also be found. The earliest entry from any of these is 1973; the update is monthly and abstracts and keywords are provided.

DAIRY SCIENCE ABSTRACTS, 1939-
Commonwealth Agricultural Bureaux, prepared by Commonwealth Bureau of Dairy Science and Technology. Quarterly 1939-1951, Monthly from 1952. Supersedes: List of References from Current Literature (National Institute for Research in Dairying).

Arrangement: Arranged by broad subject areas subdivided into smaller topics.

Coverage: Husbandry and milk production, technology, economics, physiology and biochemistry, nutrition, immunology, microbiology, chemistry and physics, and dairy research and education are covered. (See Table of Contents for subdivisions.) Annual reports, etc. are in separate sections, but the entries continue the abstract numbering.

Scope: International.

Locating Material: The abstracts are numbered consecutively through each volume. Entries in the index refer to these numbers.

Abstracts: The titles and abstracts are in English. If the article is in a language other than English, the title in English is bracketed and repeated in the original language. The language and any summaries in other languages are indicated by an abbreviation (an explanation follows the author index). The number of references is given and the address of the first author is noted when it is known.

Indexes: An author index is in each issue, and beginning with Volume 34, a subject index (keyword) is found in each issue. There are annual author and subject indexes.

Other Material: A list of abbreviations used in the abstracts is in the preliminary pages of the first issue of each volume. A list of abbreviations for language is in each issue. There is a review article in most issues. A list of liaison officers with addresses is provided at intervals. The Commonwealth Agricultural Bureaux offer a variety of services in connection with their publications. These include magnetic tapes of a consolidated database (with samples and prices available on request), search services, annotated bibliographies, and a document delivery service.

Periodicals Scanned: A list is published at intervals; normally it is found in the first issue of alternate volumes.

Database: Citations are available through CAB ABSTRACTS from 1973. These are updated monthly and contain abstracts and keywords. The other abstract journals of the Commonwealth Agricultural Bureaux are also included.

FABA BEAN ABSTRACTS, 1981-
Commonwealth Agricultural Bureaux. Quarterly.

Arrangement: Arranged by subject.

Coverage: Breeding and selection, varieties and varietal resistance, diseases and physiological disorders, climate and environment, harvesting, storage and quality, nutrition and utilization, economics, rural development and land use are among the subjects listed in the Contents.

Scope: International.

Locating Material: Entries are numbered consecutively through each volume. These numbers are referred to in the indexes.

Abstracts: Titles and abstracts are in English. If the article is in a language other than English, the language is indicated, as are available summaries. The number of references is given and the affiliation of the senior author is provided.

Indexes: Author and subject indexes are in each issue.

Other Material: A list of abbreviations for journals and for language is provided along with brief explanatory material. A description of Commonwealth Agricultural Bureaux services and products and a list of addresses and liaison officers are found.

Periodicals Scanned: A list appears at intervals.

Database: Only the main abstract journals of the Commonwealth Agricultural Bureaux are listed as being in the CAB ABSTRACTS database. However, entries in the more specialized journals such as Faba Bean Abstracts are derived from the same database so they can be found also. The earliest entry for any of these is 1973; the update is monthly and abstracts and keywords are provided.

FARM AND GARDEN INDEX, 1978-
Bell and Howell. Quarterly.

Arrangement: Arranged by subject.

Coverage: The introduction states that "...the focus of the index is on popular magazines that provide timely, practical information for farmers and gardeners." Research journals in agriculture and horticulture are also included.

Scope: American journals.

Locating Material: The index is divided into two sections. Subjects are followed by personal names.

Abstracts: Not applicable. The entries give complete title of articles, periodical volume, number, page and date. The personal names section includes authors and people cited in articles.

Other Material: Brief explanatory material is provided.

Periodicals Scanned: The list is found in each issue.

FERTILIZER ABSTRACTS, 1968-
Tennessee Valley Authority. Monthly.

Arrangement: Arranged by four broad topics.

Coverage: The topics are Analytical Methods, Technology, Marketing, and Use. In the introductory material it is stated that the journal contains "information on fertilizer technology, marketing, use, and related research."

Scope: International.

Locating Material: Entries are numbered consecutively through each volume. These numbers are referenced in the indexes.

Abstracts: Titles and abstracts are in English. If the article is in another language, the language is noted. Occasionally the abstracts are taken from Chemical Abstracts and if so, the CA volume and entry number are given. Affiliation of the senior author is noted. Occasionally no abstract is provided and sometimes they are quite long, with varying lengths between these extremes.

Agricultural Sciences / 151

Indexes: Author and subject indexes are in each issue. These cumulate annually.

FIELD CROPS ABSTRACTS, 1948-
Commonwealth Agricultural Bureaux, prepared by Commonwealth Bureau of Pastures and Field Crops. Irregular, 1948-1949, Quarterly 1950-1972, Monthly from 1973.

Arrangement: Arranged by subject. Sections and subsections for general and more specific topics.

Coverage: Specific annual cereals, legumes, root crops, oilseeds and fiber crops, crop botany, pests and diseases, surveys and land use, farming systems, soil and water conservation and agrometeorology are covered. (See Table of Contents for a complete list.) Reports, meetings and books are included (see Other Material, below).

Scope: International.

Locating Material: Entries are numbered consecutively through each volume. Index entries refer to these numbers.

Abstracts: Titles and abstracts are in English. If the article is in another language, the title is bracketed. Abbreviations indicating the language of the article and summaries in other languages are given (explanation of these is found in the preliminary pages of each issue). The number of references cited and the address of the first author is provided. The abstracts may be quite detailed or very brief. Occasionally none is provided.

Indexes: There is an author and a subject index in each issue which cumulate annually.

Other Material: A review article appears approximately quarterly. A list of abbreviations of languages and of terms used in the abstracts is found in each issue. Reports, meetings, books, "News and Notes" are found at the end of each issue. These entries continue the consecutive numbering of the abstracts. A list of libraries which hold periodicals scanned is given. Annotated bibliographies on specified subject profiles are provided. The Commonwealth Agricultural Bureaux offer a variety of services in connection with their publications. These are described in each issue.

Periodicals Scanned: A list is provided at intervals, usually in the first issue of a volume.

Database: Citations are available through CAB ABSTRACTS. They are updated monthly and abstracts and keywords are provided.

The other abstract journals of the Commonwealth Agricultural Bureaux are also included.

FOOD SCIENCE AND TECHNOLOGY ABSTRACTS, 1969-
International Food Information Service. Monthly.

Arrangement: Arranged by subject. Divided into nineteen sections of broad subject areas, each identified by a letter of the alphabet.

Coverage: Covers widely all material concerned with new developments and research in Food Science and Technology including composition, microbiology, hygiene, basic food science, toxicology, standards, legislation, engineering processing, packaging and additives, all human food commodities and aspects of food processing, and consumption of finished foods. Books and patents as well as journals are included.

Scope: International.

Locating Material: Within each section, abstracts are in random order, but each carries a unique serial number by which it may be located with the indexes. The abstract number is composed of three parts, the first being the issue number, the second the section letter, the third being the number of the abstract within its section. For example, 12 J 604 refers to the 604th article in section J in issue number 12 (December) of a given volume.

Abstracts: Titles and abstracts are in English. If the article is in another language the title will be bracketed. German articles and patents give the German titles as well as the English. All books in languages other than English have the title in the language of the book as well as in English. The language of articles and books, and summaries in other languages are indicated. All authors are given and editors are so specified. The address of the senior author is given, as are the initials of the abstractor. Abstracts range from brief to rather long.

Indexes: Author and detailed subject indexes are in each issue. These cumulate annually. "See also" and cross references are provided to facilitate use. Abstracts are indexed under the most specific headings so that the searcher with a broad subject must also check under the more specific topics.

Other Material: Explanatory material is provided with a list of abbreviations used. A list of abstractors is furnished. A description of the organization, objectives and services of International Food Information Service is given.

Periodicals Scanned: A list is in the first issue of each volume.

Database: Citations are available from 1969 through FSTA. There are monthly updates and abstracts and keywords are provided.

FOREST PRODUCTS ABSTRACTS, 1978-
Commonwealth Agricultural Bureaux, Monthly.

Arrangement: Arranged by broad subjects with subdivisions for narrower topics.

Coverage: The subjects listed in the contents include general aspects of forest products and industry; wood properties; timber extraction; damage to timber and timber protection; utilization of wood; veneers; and pulp industries.

Scope: International.

Locating Material: Entries are numbered consecutively through each volume; these numbers are referred to in the indexes.

Abstracts: Abstracts are in English. The titles are in English and the language of the article if a Latin alphabet is used. The language of the article and summaries in other languages are noted. An abbreviation for a library source is provided. Illustrative material and the number of references are indicated.

Indexes: Author and subject indexes are found in each issue (until 1983, author and species indexes). Annual author, species, and subject indexes are provided.

Other Material: Extensive explanatory material is provided which includes lists of abbreviations used. Occasionally a review article is found.

Database: Citations are available through CAB ABSTRACTS from 1978 and are updated monthly. Keywords and abstracts are provided. The other abstract journals of the Commonwealth Agricultural Bureaux are also included.

FORESTRY ABSTRACTS, 1939-
Commonwealth Agricultural Bureaux, prepared by Commonwealth Forestry Bureau. Quarterly, 1939-1972, Monthly from 1973.

Arrangement: Arranged by broad subject areas.

154 / Abstracts and Indexes

Coverage: Broad coverage, including all aspects of forestry such as silviculture, mensuration, management, physics, environment, fire, plant biology, genetics and breeding, mycology and pathology, insects and other invertebrates, range, game and wildlife, fish, protection, other land use, and dendrochronology and dendroclimatology is provided.

Scope: International.

Locating Material: Entries are numbered consecutively through each volume.

Abstracts: The titles and abstracts are in English. The titles are given in the original language if it is other than English and a Latin alphabet is used. Russian and Greek are given in standard transliteration for whole publications. The name of the language and summaries in other languages are noted. The number of references and illustrative material are given. Frequent "see also" references guide the user to related material.

Indexes: Author and subject indexes are found in each issue. Annual indexes include cumulative author index, species index (which is alphabetical by scientific name with subheadings), and subject index (alphabetical with subheadings).

Other Material: Review articles are found in some issues, and also occasional retirement notices and obituaries, advertisements of books, journals, and forestry societies (the latter giving information on objectives, membership prices, publications, etc.). Explanatory material which facilitates the use of Forestry Abstracts is found in several issues. A list of abbreviations used is provided.

Periodicals Scanned: A list is supplied at intervals.

Database: Citations are provided through CAB ABSTRACTS from 1973. The other abstract journals of the Commonwealth Agricultural Bureaux are also included in the database.

HERBAGE ABSTRACTS, 1931-
    Commonwealth Agricultural Bureaux, prepared by Commonwealth Bureau of Pastures and Field Crops. Quarterly, 1931-1972, Monthly from 1973.

Arrangement: Arranged by broad subject areas.

Coverage: Management of cultivated grasslands, rangelands, and fodder crops, herbage plant varieties, productivity, chemical composition and nutritive value for animals, agrometeorology,

Agricultural Sciences / 155

biology, physiology and ecology herbage plants, environmental contamination, etc. are found as subjects. Reports and meetings are included (see Other Material below).

Scope: International.

Locating Material: Entries are numbered consecutively through each volume. The indexes refer to these numbers.

Abstracts: Titles and abstracts are in English. If the article is in another language the title in English is bracketed and repeated in the language of the article. The language and summaries in other languages are indicated by abbreviations (a key is provided). The number of references and the address of the first author is given.

Indexes: Author and subject indexes are in each issue with annual cumulations. A cumulative index for Volumes 1-10 is available.

Other Material: A list of abbreviations and language symbols is provided in the preliminary pages of each issue. A list of common names is given in the annual index. Reports, meetings, and book reviews are carried separately at the end of each issue but the entries continue the consecutive numbering of the abstracts. Review articles appear approximately quarterly. A list of liaison officers with addresses is included at intervals.

Periodicals Scanned: A list is published at intervals, usually in the first issue of a volume.

Database: Citations are available through CAB ABSTRACTS from 1973. These are updated monthly and contain abstracts and keywords. The other abstract journals of the Commonwealth Agricultural Bureaux are also included.

HOME ECONOMICS RESEARCH ABSTRACTS, 1966-
American Home Economics Association. Annual. Published in the following sections.
    Art and Housing, Furnishings and Equipment, 1967-
    Family Economics, Home Management, 1966-
        Supersedes: Journal of Family Economics-Home Management, Volumes 1-5, 1962-66.
    Family Relations and Child Development, 1966-
    Food and Nutrition, 1968-
        Communications included in 1971.
    Institution Administration, 1967-
    Textiles and Clothing, 1966-

Arrangement: Arrangement of sections varies slightly. Usual arrangement is under broad subject headings (within the scope of the section), subdivided into smaller groups, then alphabetical by author.

Coverage: Covers fields in Home Economics as listed above. (See Table of Contents for each section.)

Scope: Limited to abstracts of masters' theses and doctoral dissertations completed in graduate schools of home economics in colleges and universities in the United States. Also depends upon the number of institutions submitting answers to the request for their lists.

Locating Material: References from the index are to the page number on which the citation appears.

Abstracts: These are usually informative. Complete bibliographic data and such information as type of degree (PH.D or Master's), date completed, university issuing the degree and whether or not the thesis or dissertation is available on Interlibrary Loan is given.

Indexes: An author index is provided which gives the author's name and full title.

Other Material: The "Introduction" explains how to obtain further information and how many institutions submitted abstracts. Ordering information is in the back of the issue.

HORTICULTURAL ABSTRACTS, 1931-
Commonwealth Agricultural Bureaux, prepared by Commonwealth Bureau of Horticulture and Plantation Crops. Quarterly, 1931-1972, Monthly from 1973.

Arrangement: Arranged by subject. Sections and subsections for general and more specific areas. (See Contents in each issue for comprehensive list.)

Coverage: Research and its application, tree fruits and nuts, temperate tree fruits and nuts, small fruits and vines, vegetables, ornamental plants, minor temperate and tropical industrial crops, subtropical fruit and plantation crops, tropical fruit and plantation crops (broad headings are subdivided by species and aspects of management). Books, symposia, conference proceedings and annual reports are included.

Scope: International.

Locating Material: Entries are numbered consecutively through each volume. These numbers are referred to in the indexes.

Abstracts: The titles and abstracts are in English. If the article
is in another language, the title is bracketed and if it is in
Roman script, it is repeated in the language of the article.
An abbreviation indicating the language and summaries in
other languages is provided. The name of the journal is
given in full and all authors' names and the affiliation of the
senior author are given. Number of references and illustrative
material are indicated. The abstracts are usually concise
but informative. "See also" references at the beginning of
subject sections lead the searcher to related items. This
description refers to the journal as it presently appears. In
earlier volumes the searcher will encounter such minor differences
as abbreviated journal titles and languages spelled
out. Occasionally no abstract will be provided (if the title
is self-explanatory or if the subject is marginal).

Indexes: Author and subject indexes are provided with each issue.
These cumulate annually. There are cumulative author and
subject indexes to Volumes 1-10, 11-15, 16-20, 21-25, 26-30.

Other Material: A review article is in some issues (beginning with
Volume 42). A key to abbreviations used in the entries and
in the abstracts and a list of recent annotated bibliographies
are found in the preliminary pages of each issue. There is
a description of Commonwealth Agricultural Bureaux products
and services.

Periodicals Scanned: A list is published at regular intervals, usually
in the first issue of a volume.

Database: Citations are available through CAB ABSTRACTS from
1973. These are updated monthly and contain abstracts and
keywords. The other abstract journals of the Commonwealth
Agricultural Bureaux are also included.

IRRIGATION AND DRAINAGE ABSTRACTS, 1975-
Commonwealth Agricultural Bureaux. Quarterly.

Arrangement: Arranged by broad subjects most of which have subtopics.

Coverage: Water management, irrigation systems and requirements
and irrigation of croplands (with an extensive list of crops
covered), drainage, soil/water relations, plant/water relations,
salinity and toxicity problems, meteorological aspects,
environmental and public health aspects are covered.

Scope: International.

Locating Material: Entries are numbered consecutively through each
volume. These numbers are referred to in the indexes.

Abstracts: Titles and abstracts are in English. The language of the article is indicated, as are summaries in other languages. The number of references is given and the affiliation of the senior author is provided.

Indexes: There is an author index in each issue which cumulates annually.

Other Material: A list of abbreviations for journals and for language is provided with brief explanatory material. A description of Commonwealth Agricultural Bureaux services and products, a list of addresses and liaison officers are provided.

Periodicals Scanned: A list appears at intervals.

Database: Only the main abstract journals of the Commonwealth Agricultural Bureaux are listed as being in the CAB ABSTRACTS database. However, entries in the more specialized journals such as Irrigation and Drainage Abstracts are derived through the same database so they can also be found. The earliest entry for any of these is 1973; the update is monthly and abstracts and keywords are provided.

MAIZE QUALITY AND PROTEIN ABSTRACTS, 1975-
Commonwealth Agricultural Bureaux. Quarterly.

Arrangement: Arranged by subject.

Coverage: Comprehensive information on the subject is provided from a wide variety of literature. This journal is an example of abstracts collected from the CAB database on a "specialized topic of wide interest."

Scope: International.

Locating Material: Entries are numbered consecutively through each volume.

Abstracts: Titles and abstracts are in English. If the article is in a language other than English the language and any summaries available are indicated. The number of references is noted and the affiliation of the senior author is given.

Indexes: Author and subject indexes are found in each issue.

Other Material: A list of abbreviations for journals and for languages is provided along with brief explanatory material in the preliminary pages. A description of Commonwealth Agricultural Bureaux services and products, a list of addresses, and a list of liaison officers are found.

Periodicals Scanned: A list appears occasionally.

Database: Only the main abstract journals of the Commonwealth Agricultural Bureaux are listed as being in the CAB ABSTRACTS database. However, entries from the more specialized journals such as Maize Quality and Protein Abstracts are derived through the same database so they can also be found. The earliest entry for any of these is 1973; the update is monthly and abstracts and keywords are provided.

NUTRITION ABSTRACTS AND REVIEWS, 1931-
Commonwealth Agricultural Bureaux, prepared by Commonwealth Bureau of Nutrition. Series A, Human and Experimental; Series B, Livestock Feeds and Feeding (Split into the two series with V. 47, 1977). Quarterly to 1972; Monthly from 1973.

Arrangement: Arranged by broad subjects subdivided into narrower topics.

Coverage: Series A: The broad subjects covered are foods, physiology and biochemistry, human health and nutrition, disease and therapeutic nutrition; Series B: Technique, technology, feedingstuff, and feeds, physiology and biochemistry, feeding of animals, diet in etiology of disease, proceedings, symposia, book reviews, and reports are included.

Scope: International.

Locating Material: The entries are numbered consecutively through each volume. These numbers are referred to in the indexes.

Abstracts: The titles are given in the language of the article and in English. The language and any summaries in another language are noted by an abbreviation. The senior author's affiliation is given. Generally the abstracts are complete, giving detailed information. Occasionally there will be no abstract, citation by author and title only. The abstracts are in English, and the number of references is stated.

Indexes: Each issue carries an author and subject index which cumulates annually. An annual combined Table of Contents is provided.

Other Material: Review articles are found in some issues. Abbreviations and acronyms used are explained in the preliminary pages of each issue. A list of liaison officers with addresses is provided once a year. A description of Commonwealth Agricultural Bureaux services and products is given.

160 / Abstracts and Indexes

Database: Citations are available through CAB ABSTRACTS. They are updated monthly and abstracts and keywords are provided. The other abstract journals of the Commonwealth Agricultural Bureaux are included.

NUTRITION PLANNING, 1978-
Oelgeschlager, Gunn and Hain, Publishers, Inc. Quarterly.

Arrangement: Arranged by ten subject categories, then alphabetically within each category.

Coverage: The ten categories are: planning process, methodology, and analysis; consequences of malnutrition; nutritional status assessment; nutrition education and home centered activities; public health and curative measures; food processing, distribution, and feeding programs; agriculture; economics; social and cultural aspects; comprehensive programs. The Table of Contents lists the subjects included in these categories and indicates the document number, short title and frequently the author's last name.

Scope: International.

Locating Material: The entries are numbered consecutively through each journal and through successive journal issues. These numbers are referred to as Document Numbers and are used to identify the documents or abstracts.

Abstracts: The entries include the bibliographic citation, the abstract, and the full-text availability information. The abstract appears in only the most appropriate category.

Indexes: Geographic, source (corporate and personal authors), and a subject index are provided in each issue. These cumulate annually.

Other Material: There is detailed descriptive material provided; a list of participating organizations, book reviews, and a newsletter section are planned.

Periodicals Scanned: As sources of information, these would appear in the Source Index.

ORNAMENTAL HORTICULTURE, 1975-
Commonwealth Agricultural Bureaux. Monthly.

Arrangement: Arranged by subject.

Agricultural Sciences / 161

Coverage: Annual and herbaceous plants, bulbs, and tubers, lawns and turf, aquatic plants, bromeliads, foliage and succulent plants, and trees and shrubs are included. Reports, conferences, and books are also covered.

Scope: International.

Locating Material: Entries are numbered consecutively through each volume. Index entries refer to these numbers.

Abstracts: The abstracts are in English. The titles are in English and usually in the language of the article. The language and summaries in translation are indicated. The number of references is noted if this is significant and the senior author's affiliation is provided.

Indexes: Author and subject indexes are found in each issue.

Other Material: A list of abbreviations for languages and for journals is provided with brief explanatory material. A description of Commonwealth Agricultural Bureaux services and products and a list of addresses and liaison officers are provided.

Database: Only the main abstract journals of the Commonwealth Agricultural Bureaux are listed as being in the CAB ABSTRACTS database. However, entries in the more specialized journals such as Ornamental Horticulture are derived through the same database, so they can also be found. The earliest entry from any of these is 1973; the update is monthly and abstracts and keywords are provided.

PIG NEWS AND INFORMATION, 1980-
Commonwealth Agricultural Bureaux, Quarterly.

Arrangement: Arranged by subject, some of which have subdivisions.

Coverage: Behaviour, breeding, diseases, economics, housing environment and equipment, husbandry, immunology, nutrition, pharmacology, physiology and biochemistry, general pig production, pigs in biomedical research, and reproduction are subjects included in the Table of Contents.

Scope: International.

Locating Material: Entries are numbered consecutively through each volume. These numbers are referred to in the indexes.

Abstracts: The abstracts are in English. Titles are in English and in the language of the article and summaries in other languages are noted by an abbreviation. Illustrative material and the number of references are indicated. The address of the senior author is provided.

Indexes: Author and subject indexes are in each issue and cumulate annually.

Other Material: News Items, review articles and a country report are found in each issue. A list of abbreviations for journals and for languages is provided with brief explanatory material. A description of Commonwealth Agricultural Bureaux services and products, a list of addresses, and liaison officers are provided.

Database: Only the main abstract journals of the Commonwealth Agricultural Bureaux are listed as being in the CAB ABSTRACTS database. However, the more specialized journals such as Pig News and Information are derived through the same database, so they can be found also. The earliest entry for any of these is 1973; the update is monthly and abstracts and keywords are provided.

PLANT BREEDING ABSTRACTS, 1930-
Commonwealth Agricultural Bureaux, prepared by Commonwealth Bureau of Plant Breeding and Genetics. Quarterly.

Arrangement: Arranged by broad subject areas subdivided into smaller topics.

Coverage: Genetics, cytology, crop plants, grasses, roots and tubers, fibers, sugar and starch plants, stimulants, fruits and nuts, forest trees (from 1982 the material in the "Forest Tree" section is not covered in this journal, but in Forestry Abstracts) and vegetables. (See Table of Contents for a complete list.)

Scope: International.

Locating Material: Entries are numbered consecutively through each volume. These numbers are referred to in the indexes.

Abstracts: All authors are named and the senior author's affiliation is given. Titles are in English and in the language of the article, and the abstracts are in English. If the article is in a language other than English, the English version of the title is bracketed. The language of the article and summaries in other languages are indicated by an abbreviation. The number of references is noted. If the abstract is from another source, that source is cited. Beginning with Volume 42 (1972) the titles of journals are printed in full. This description pertains to the journal as it presently appears. The searcher will notice minor differences in volumes prior to 1972, e.g., the titles of journals are abbreviated and the language of articles is spelled out.

Indexes: Author and subject indexes are in each issue and cumulate annually.

Other Material: A key to abbreviations for languages is provided in the preliminary pages of each issue. Book reviews are grouped together at the end of an issue, but continue the consecutive numbering of the abstracts. Occasionally a list of new journals in the field will be provided. Commonwealth Agricultural Bureaux services and products are described.

Periodicals Scanned: A list is provided at intervals, usually in the first issue of a volume.

Database: Citations are available through CAB ABSTRACTS from 1973. These are updated monthly and contain abstracts and keywords. The other abstract journals of the Commonwealth Agricultural Bureaux are also included.

PLANT GROWTH REGULATOR ABSTRACTS, 1975-
Commonwealth Agricultural Bureaux. Monthly.

Arrangement: Arranged by broad subject areas subdivided into smaller topics.

Coverage: Auxins, gibberillins, cytokinins, ethylene and ethelene releasers, growth inhibitors, growth retardants, gametocides are covered. (See Table of Contents for a complete list of subdivisions.)

Scope: International.

Locating Material: Entries are numbered consecutively through the volume. These numbers are referred to in the indexes.

Abstracts: Titles are in English and in the language of the article. If the article is in a language other than English, the English version of the title is bracketed and the language of the articles and summaries in other languages is indicated by an abbreviation. The number of references is noted. All authors are named and the affiliation of the senior author is given. The abstracts are in English.

Indexes: There is an author index in each issue which cumulates annually. An annual subject index is provided.

Other Material: A key to abbreviations used and for languages is found in the preliminary pages of each issue. This journal is an example of the periodic collection of abstracts on specialized topics drawn from the whole CAB database. Commonwealth Agricultural Bureaux services and products are described.

164 / Abstracts and Indexes

Database: Only the main abstract journals of the Commonwealth Agricultural Bureaux are listed as being in the CAB ABSTRACTS database. However, entries in the more specialized journals such as Plant Growth Regulator Abstracts are derived through the same database, so these entries can be found also. The earliest entry from any of these is 1973; the update is monthly and abstracts and keywords are provided.

POTATO ABSTRACTS, 1976-
Commonwealth Agricultural Bureaux. Monthly.

Arrangement: Arranged by subject.

Coverage: Breeding and selection, varieties and varietal resistance, agronomy, fertilizers, weeds, pests, diseases and physiological disorders, climate and environment, growth, physiology and biochemistry, harvesting, storage and quality, food technology, nutrition and utilization, economics, rural development and land use are listed in the Contents as the subjects covered. Reports, meetings, and books are also included.

Scope: International.

Locating Material: Entries are numbered consecutively through each volume. These numbers are referred to in the indexes.

Abstracts: Titles and abstracts are in English. The language of the article is indicated, as are available summaries in other languages. The number of references is given and the affiliation of the senior author is provided.

Indexes: There is an author index in each issue which cumulates annually.

Other Material: A list of abbreviations for journals and for languages is provided along wiht brief explanatory material. A description of Commonwealth Agricultural Bureaux services and products, a list of addresses, and liaison officers are provided.

Periodicals Scanned: A list appears at intervals.

Database: Only the main abstract journals of the Commonwealth Agricultural Bureaux are listed as being in the CAB ABSTRACTS database. However, entries in the more specialized journals such as Potato Abstracts are derived through the same database so they can also be found. The earliest entry for any of these is 1973; the update is monthly and abstracts and keywords are provided.

Agricultural Sciences / 165

POULTRY ABSTRACTS, 1975-
Commonwealth Agricultural Bureaux. Monthly.

Arrangement: Arranged by broad subjects some of which have subtopics.

Coverage: Fowls in general and their growth and meat production, egg production, incubation and hatching, reproduction and artificial insemination, genetics and progeny, performance tests and the evaluation of feedingstuffs, turkeys, ducks, geese and a general section which includes housing and equipment, poultry products, anatomy, physiology and biochemistry, immunology and infectious diseases, parasitic disorders, and nutritional and metabolic disorders are found. Reports, conferences, and books are also covered.

Scope: International.

Locating Material: Entries are numbered consecutively through each volume. These numbers are referred to in the indexes.

Abstracts: Titles and abstracts are in English. The language of the article is indicated as are summaries available in other languages. The number of references is given and the affiliation of the senior author is provided.

Indexes: Author and subject indexes are in each issue. These cumulate annually.

Other Material: A list of abbreviations for journals and for languages is provided with brief explanatory material. A description of Commonwealth Agricultural Bureaux services and products and list of addresses and liaison officers are provided.

Periodicals Scanned: A list appears at intervals.

Database: Only the main abstract journals of the Commonwealth Agricultural Bureaux are listed as being in the CAB ABSTRACTS database. However, entries in the more specialized journals such as Poultry Abstracts are derived through the same database, so they can be found also. The earliest entry from any of these is 1973; the update is monthly and abstracts and keywords are provided.

REVIEW OF APPLIED ENTOMOLOGY, SERIES A, AGRICULTURAL
see REVIEW OF APPLIED ENTOMOLOGY

RICE ABSTRACTS, 1978-
   Commonwealth Agricultural Bureaux. Monthly.

Arrangement: Arranged by subject.

Coverage: Breeding and selection, varieties and varietal resistance, agronomy, fertilizers, weeds, pests, diseases and physiological disorders, climate and environment, growth, physiology and biochemistry, harvesting and storage, food technology, nutrition and utilization, economics, rural development and land use are listed in the Table of Contents. Books, reports, and meetings are included.

Scope: International.

Locating Material: Entries are numbered consecutively through each volume. These numbers are referred to in the indexes.

Abstracts: Abstracts are in English. Titles are in English and usually in the language of the article. The language and summaries in other languages are indicated. The number of references is noted.

Indexes: An author index is in each issue which cumulates annually. There is an annual subject index.

Other Material: A list of abbreviations for languages and for journals is provided with brief explanatory material. A description of Commonwealth Agricultural Bureaux services and products, a list of addresses and liaison officers is provided.

Database: Only the main abstract journals of the Commonwealth Agricultural Bureaux are listed as being in the CAB ABSTRACTS database. However, entries in the specialized journals such as Rice Abstracts are derived through the same database, so they can also be found. The earliest entry for any of these is 1973; the update is monthly and abstracts and keywords are provided.

SEED ABSTRACTS, 1978-
   Commonwealth Agricultural Bureaux. Monthly.

Arrangement: Arranged by subject.

Coverage: Seed morphology and anatomy, physiology and biochemistry of seed development, seed chemistry, production, storage and longevity, germination, pests and diseases, breeding and selection, processing, assessment and testing, and economics and marketing are among the subjects listed in the Contents. Books, conferences, and meetings are also included.

Scope: International.

Locating Material: Entries are numbered consecutively through each volume. Index entries refer to these numbers.

Abstracts: Abstracts are in English. Titles are in English and usually in the language of the article. The language and summaries in other languages are indicated. The number of references is noted if this is significant.

Indexes: There is an author and species index in each issue.

Other Material: A list of abbreviations for languages and for journals is provided with brief explanatory material. A description of Commonwealth Agricultural Bureaux services and products and list of addresses and liaison officers are provided. Occasionally a review article will be found.

Database: Only the main abstract journals of the Commonwealth Agricultural Bureaux are listed as being in the CAB ABSTRACTS database. However, entries in the more specialtized journals such as Seed Abstracts are derived through the same database, so they can be found also. The earliest entry for any of these is 1973; the update is monthly and abstracts and keywords are provided.

SOILS AND FERTILIZERS: 1938-
Commonwealth Agricultural Bureaux, prepared by Commonwealth Bureau of Soils. Six issues a year through 1972. Monthly from 1973.

Arrangement: Arranged by subject according to a decimal classification (key provided in the Contents). Broad sections are divided into subsections for more specific topics.

Coverage: Soils, soil/plant relationships, mineralogy, experimentation, physical properties of soils, soil classification, use of fertilizers, plant nutrition, plant diseases and protection, plant pests, crop culture, fruit, forestry, horticultural crops, earth sciences, botany, and ecology are included. (See Table of Contents for a complete list.)

Scope: International.

Locating Material: The entries are numbered consecutively through each volume. The indexes refer to these numbers.

Abstracts: Titles and abstracts are in English. If the article is in another language, the title is bracketed. The language and any summaries in other languages are indicated by an abbreviation. The address of the first author is given. Beginning with Volume 35, titles of journals were no longer abbreviated. Abstracts are sometimes quite brief, but usually detailed enough to give good information. Occasionally none appears.

Indexes: Author and subject indexes are in each issue. These cumulate annually.

Other Material: Review articles are found in some issues. A list of symbols for languages is in each issue. A book list is found in a separate section following the abstracts of journal articles. Occasionally there will be a list of annotated bibliographies available from the Commonwealth Bureau of Soils. A list of liaison officers with addresses is provided at intervals. The Commonwealth Agricultural Bureaux services and products are described.

Periodicals Scanned: A list is usually provided in the first issue of a volume but not always (none found in Volume 45, number 1).

Database: Citations are available through CAB ABSTRACTS. They are updated monthly and abstracts and keywords are provided. The other abstract journals of the Commonwealth Agricultural Bureaux are also included.

SORGHUM AND MILLETS ABSTRACTS, 1976-
Commonwealth Agricultural Bureaux. Monthly.

Arrangement: Arranged by subject.

Coverage: Economics, breeding and varieties, planting and germination, soils, nutrition and management, diseases and pests, physiology, biochemistry and growth, harvesting and storage are among the aspects of sorghum (including hybrids) which are treated. Finger, Common and Pearl millet, minor millets and related crops are also included.

Scope: International.

Locating Material: Entries are numbered consecutively through each volume. These numbers are referred to in the indexes.

Abstracts: Titles and abstracts are in English. Journal titles are given in full and if the article is in a language other than English, that language and summaries in other languages. are indicated. The number of references is given as well as the address of the first author.

Indexes: Author and subject indexes are found in each issue.

Other Material: A list of abbreviations for languages and terms used in the abstracts is found along with brief explanatory material in the preliminary pages.

Periodicals Scanned: A list appears at intervals.

Database: Only the main abstract journals of the Commonwealth Agricultural Bureaux are listed as being in the CAB ABSTRACTS database. However, entries in the more specialized journals such as Sorghum and Millets Abstracts are derived through the same database so they can also be found. The earliest entry for any of these is 1973; the update is monthly and abstracts and keywords are provided.

SOYABEAN ABSTRACTS, 1978-
Commonwealth Agricultural Bureaux. Monthly.

Arrangement: Arranged by subject.

Coverage: Breeding and selection, varieties and varietal resistance, agronomy, weeds, pests, diseases and physiological disorders, climate and environment, nitrogen fixation, growth, physiology and biochemistry, harvesting, storage and quality, food technology, nutrition and utilization, economics, rural development and land use are listed in the Table of Contents. Reports, meetings, and books are also covered.

Scope: International.

Locating Material: Entries are numbered consecutively through each volume. Indexes refer to these numbers.

Abstracts: Abstracts are in English. Titles are in English and usually in the language of the article. The language and summaries in other languages are indicated. The number of references is noted if this is significant.

Indexes: An author index is in each issue which cumulates annually. There is an annual subject index.

Other Material: A list of abbreviations for languages and for journals is provided with brief explanatory material. A description of Commonwealth Agricultural Bureaux services and products and list of addresses and liaison officers are provided.

Database: Only the main abstract journals of the Commonwealth Agricultural Bureaux are listed as being in the CAB ABSTRACTS database. However, entries in the more specialized journals such as Soyabean Abstracts are derived through the same database, so they can also be found. The earliest entry of any of these is 1973; the update is monthly and abstracts and keywords are provided.

TEXTILE TECHNOLOGY DIGEST, 1944-
Institute of Textile Technology. Monthly.

Arrangement: Arranged by broad subject areas subdivided into smaller topics.

Coverage: All aspects of textile technology including yarn, fabric, fibers, and apparel; production finishing; testing; mill management; the chemistry, physics, and biology of fibers, yarns, and fabrics; and general information on research are covered. Patents are included, but are separate from journal articles and books.

Scope: International.

Locating Material: Entries are numbered consecutively through each volume. (The two digits at the end of the number indicate the year.)

Abstracts: Brief but informative abstracts with full bibliographical information are provided. Titles and abstracts are in English and the original language of the article is noted.

Indexes: An author index is in each issue. In 1972 cumulative six-month and annual indexes were published, consisting of author, patent concordance, and subjects. The patent index is a numerical listing by country. The patent concordance links together patents from different countries which correspond to one another, i.e., cover the same invention. It also includes entries for newly discovered patents which correspond to earlier patents. Preceding the patent concordance is an explanation of its use.

Other Material: Names and addresses of officers, trustees, and members of the Technical Advisory Committee are given and occasionally a list of subject headings will be provided.

Periodicals Scanned: The list used to appear in the first issue of the year but is now provided irregularly.

TOBACCO ABSTRACTS, 1957-
Tobacco Literature Service, North Carolina State University. Frequency varies (monthly, bimonthly).

Arrangement: Arranged by subject, then alphabetical by author within each subject.

Coverage: Chemical and physical properties, climatological factors, various bases of diseases (bacteria, fungus, nematode, virus), the genetics and varieties, and harvesting and curing of tobacco, and its physiology and biochemistry are covered. Economics, marketing, production, and manufacturing technology are also included. State Agricultural Experiment Station publications, federal documents, and patents, are also provided.

Scope: International.

Locating Material: The abstracts are numbered consecutively through each volume. These numbers are referred to in the indexes.

Abstracts: Titles and abstracts are in English. The abstracts vary in length from half a page to one line. Occasionally none appears. Abbreviations are used instead of full journal titles. If the article is in a language other than English, an abbreviation is used to indicate that language. Summaries in English are also noted.

Indexes: Annual author and subject indexes are provided.

Periodicals Scanned: No list is provided, but journal titles are frequently found in the author index.

TRITICALE ABSTRACTS, 1975-
Commonwealth Agricultural Bureaux. Quarterly.

Arrangement: Arranged by subject.

Coverage: Comprehensive information on the subject is provided from a wide variety of literature. This journal is an example of abstracts collected from the CAB database on a "specialized topic of wide interest."

Scope: International.

Locating Material: Entries are numbered consecutively through each volume.

Abstracts: Titles and abstracts are in English. If the article is in a language other than English, the language and any available summaries are noted. The number of references is indicated and the affiliation of the senior author is given.

Indexes: Author and subject indexes are found in each issue.

Other Material: A list of abbreviations for journals and for languages is provided along with brief explanatory notes in the preliminary pages. A description of Commonwealth Agricultural Bureaux services and products, a list of addresses, and liaison officers are found.

Periodicals Scanned: A list appears occasionally.

Database: Only the main abstract journals of the Commonwealth Agricultural Bureaux are listed as being in the CAB ABSTRACTS database. However, entries in the more specialized journals such as Triticale Abstracts are derived from the same database so they can also be found. The earliest entry for any of these is 1973; the update is monthly and abstracts and keywords are provided.

TROPICAL ABSTRACTS see ABSTRACTS ON TROPICAL AGRICULTURE

TROPICAL OIL SEEDS ABSTRACTS, 1975-
Commonwealth Agricultural Bureaux. Monthly.

Arrangement: Arranged by type of plant, i.e., ground nut, safflower, coconut, oil palm, castor, and sesame. Then other oil seed crops, and by-products, and unspecified seed oils.

Coverage: In addition to the crops listed above, breeding, seed treatment, disease and pests, physiology, harvesting and by-products are covered. Books, reports, and conferences are included.

Scope: International.

Locating Material: Entries are numbered consecutively through each volume. These numbers are referred to in the indexes.

Abstracts: Titles and abstracts are in English. If the article is in a language other than English, the language is indicated, as are available summaries. The number of references is given and the affiliation of the senior author is provided.

Indexes: Author and subject indexes are found in each issue. These cumulate annually.

Other Material: A list of abbreviations for journals and for languages is provided along with brief explanatory material. A description of Commonwealth Agricultural Bureaux services and products, a list of addresses and liaison officers are found.

Periodicals Scanned: A list appears at intervals.

Database: Only the main abstract journals of the Commonwealth Agricultural Bureaux are listed as being in the CAB ABSTRACTS database. However, entries in the more specialized journals such as Tropical Oil Seeds Abstracts are derived through the same database so they can be found also. The earliest entry for any of these is 1973; the update is monthly and abstracts and keywords are provided.

WEED ABSTRACTS, 1952-
  Commonwealth Agricultural Bureaux, prepared by Weed Research Organization of the Agriculture Research Council. Monthly 1952-1962; bimonthly, 1962-1972, monthly from 1973.

Arrangement: Arranged by broad subjects subdivided into smaller topics.

Coverage: Weed control in annual field crops, grassland and herbage, vegetable crops, ornamentals, fruit crops, plantation crops, forests, aquatic weeds, weed biology, the physiology and biochemistry of herbicides, their physical and chemical properties, methods of analysis, toxicology and interaction with soil and soil micro-organisms are covered. Also the methods used to apply herbicides in crops and forestry are included.

Scope: International.

Locating Material: Entries are numbered consecutively through each volume. These numbers are referred to in the indexes.

Abstracts: Titles and abstracts are in English. If the article is in another language, the title is bracketed. Journal titles are given in full and the language of the article and summaries in other languages are indicated by an abbreviation. The number of references is given as well as the address of the first author. If the abstract is taken from another source, that source is cited. Abstracts are usually quite long and detailed. "See also" references at the beginning of subsections lead the searcher to related items.

Indexes: Author, subject, and species indexes are found in each issue. These cumulate annually.

Other Material: A list of abbreviations for languages and terms used in the abstracts is in the preliminary pages of each issue. A list of abbreviations and common names of herbicides are in the end pages. A list of addresses and liaison officers is provided. A description of Commonwealth Agricultural Bureaux services and products is found.

Periodicals Scanned: A list appears regularly, usually in the first issue of a volume.

Database: Citations are available through CAB ABSTRACTS from 1973. These are updated monthly and contain abstracts and keywords. The other abstract journals of the Commonwealth Agricultural Bureaux are also included.

WORLD AGRICULTURAL ECONOMICS AND RURAL SOCIOLOGY ABSTRACTS, 1959-

Commonwealth Agricultural Bureaux, prepared by Commonwealth Bureau of Agricultural Economics. Quarterly, 1959-1972, monthly from 1973. Preliminary issue titled: World Agricultural Economics Abstracts (July 1958).

Arrangement: Arranged by broad subject areas subdivided into smaller topics. Generally, in subsections, papers are arranged alphabetically first by region and country and where applicable by product and then by author.

Coverage: Policy, products, marketing and distribution, international trade, economics of production, etc. are covered (see Table of Contents).

Scope: International.

Locating Material: Entries are numbered consecutively through each volume. All indexes refer to these numbers.

Abstracts: Titles are given in the language of the article and in English. Abstracts are in English and usually detailed, but sometimes quite brief. Occasionally no abstract appears and the citation is by author and title only. The language and any summaries in other languages are indicated by an abbreviation. The number of references and illustrative material are noted. "See" and "see also" references at the beginning of the subject sections lead the reader to alternate terms. The author's affiliation is given when possible.

Indexes: Author and subject indexes are in each issue. These cumulate annually. Place names are included as subjects.

Other Material: Review articles are found in some issues. There is an annual list of outside contributors and sources in the final issue of each volume. A list of abbreviations for languages and for location of papers cited is provided in each issue. Names of liaison officers and their addresses are provided at intervals. A "Guide to Aid Readers" is provided in each issue beginning with Volume 24 (1982). The services and products of the Commonwealth Agricultural Bureaux are described.

Database: Citations are available through CAB ABSTRACTS from 1973. They are updated monthly and contain abstracts and keywords. The other abstract journals of the Commonwealth Agricultural Bureaux are also included.

WORLD TEXTILE ABSTRACTS, 1969-
The Shirley Institute. Semimonthly. The editor states that this publication "...continues Textile Abstracts, which was published by the Textile Institute and appeared as such from 1967-1969, though it continued the Abstracts Section of the Journal of the Textile Institute (1922-1966/67) and had been preceded by Classified Abstracts in the Journal of the Textile Institute (1917-1921). The whole serial incorporated abstracts from the Shirley Institute Summary of Current Literature (1921-1968), which composed about 90 percent of the publication; the typeset material was used from 1929 and the Summary itself was used from 1964-1965, when ... the other sections were on paper of different colors."

Arrangement: Arranged by broad subject headings subdivided into smaller topics. Broad areas are indicated by arabic numbers, smaller topics by letters of the alphabet.

Coverage: All aspects of fibers, yarns, fabrics, chemical and finishing processes, analysis, grading, testing, defects, laundering, dry cleaning, related scientific and engineering aspects of textiles and the textile industry, as well as sociological, medical, educational aspects, and relevant general material are included. British and U.S. Patents are found but are at the end of each subject section. Entries continue the consecutive numbering of other citations. Conferences and trade journals as well as scholarly journals are covered.

Scope: International.

Locating Material: Entries are numbered consecutively through the volume, with the date of the volume preceding the number. The digits indicating year are not used in the index. This number may be followed by the letter P, indicating that the citation is a patent.

Abstracts: A complete bibliographic citation, with title in English, followed by an indication of the language of the article, if other than English, is found. At the end of each abstract a capital letter, (C), (H), (L), or (W), (as explained below under Periodicals Scanned) appears. Full titles of periodicals are used in the references. Abstracts range in length from brief to quite long.

Indexes: Annual author, subject and numerical patent indexes are provided. Monthly indexes are available but must be purchased separately. Preceding the author and subject indexes is an explanation of their arrangement.

Other Material: Corrigenda appear in the index issue. Instructions for obtaining copies of items abstracted appear in each issue and in the index issue. A list of agencies collaborating in the production of the abstracts is on the back cover of each issue. An outline of the Classification appears in the first issue of each year.

Periodicals Scanned: A list appears in the first issue of each volume which gives the names of the periodicals (excluding annuals but including some trade journals) received by the libraries of the contributing research associations, and which are identified by capital letters in brackets, as follows:
   C   Cotton Silk and Man-made Fibres Research Association
   H   Hosiery and Allied Trades Research Association
   W   Wool Industries Research Association

Database: Citations are available through WORLD TEXTILES from 1970 with monthly updates. Abstracts and keywords are provided.

HEALTH SCIENCES

ABSTRACTS OF HEALTH CARE MANAGEMENT STUDIES, V. 15, September 1978-   . Health Administration Press for the Cooperative Information Center for Health Care Management Studies, The University of Michigan. Annual. Formerly: Abstracts of Hospital Management Studies, 1964-1978.

Arrangement: Arranged by subject.

Coverage: Covers "...studies of management, planning and public policy related to the delivery of health care." Such topics as nursing, manpower, insurance and prepayment, personnel, pharmacy, architecture, and administration are included in the forty-six subject categories.

Scope: International.

Locating Material: An alphanumerical system is employed for the entries. There is a five-digit number followed by letters which indicate the subject. (The example given is: 12345 PH which means study number 12345 is the subject Pharmacy.) These numbers are sequential but not consecutive. The indexes lead the user to the proper section and number.

Abstracts: Titles and abstracts are in English. The number of pages is noted at the beginning of each entry and abbreviations provide other information about the publication (e.g. "B" indicates a bibliography, "R" means an unpublished report, etc.).

Indexes: Author, subject, and microfilm indexes are provided.

Other Material: Detailed explanatory information is provided and a list of the abbreviations employed is given. Addresses for frequently cited United States or foreign government documents, and other sources are given at the end of the journal list.

Periodicals Scanned: A list with publishers' addresses of journals routinely searched is in each issue.

ABSTRACTS OF WORLD MEDICINE, 1947-1971.
British Medical Association. Ceased.

Arrangement: Arranged by broad subjects subdivided into narrower topics.

Coverage: Selective coverage of all areas of medicine including public health and industrial medicine, pharmacology, forensic medicine and toxicology.

Scope: International.

Locating Material: No numbering system is used. Reference is to the page on which the abstract appears.

Abstracts: Titles and abstracts are in English. Language of the article is noted. Abstractor's name is given or if an author's summary is used, this is noted. Illustrative material is indicated.

Indexes: Author and subject indexes are in each issue. These cumulate annually.

Other Material: A list of abbreviations is found in each issue. Frequently a review article appears.

ABSTRACTS ON HYGIENE AND COMMUNICABLE DISEASES, 1926-
Bureau of Hygiene and Tropical Diseases. Monthly. (Incorporating the Bulletin of Hygiene first published in 1926). Formerly: Abstracts on Hygiene.

Arrangement: Arranged by broad subjects subdivided into more narrow topics.

Coverage: Environmental Health including air and atmosphere, radiation, sanitation, water, food, toxicology, venoms and antivenoms, and occupational medicine; community health, which covers vital statistics, health services, family health and drug abuse; diseases and their control which incorporates communicable and noncommunicable diseases and therapeutic agents are covered.

Scope: International.

Locating Material: Entries are numbered consecutively through each volume.

Abstracts: Abstracts are in English. Titles are in English and in the language of the article. Summaries in other languages are indicated and the number of references is given if this is important.

Health Sciences / 179

Indexes: Author and subject indexes are in each issue. These cumulate annually.

Other Material: A list of contributors is provided and symbols which are used for languages and currency are explained.

Periodicals Scanned: A list is found in the January 1983 issue.

ACCUMULATIVE VETERINARY INDEX, 1960-1980.
Medical Media Corporation. Annual (supplements) with five-year cumulations; ceased.

Arrangement: The index is in two sections; the first is arranged by subject, the second by author.

Coverage: Diseases, conditions, treatment, surgical procedures, drugs, fertility, parasites, care of mammals, laboratory animals, poisoning, and body organs are included.

Scope: Entries are from American literature.

Locating Material: Searcher uses either an author or a subject approach.

Abstracts: Not applicable. Entries include complete bibliographic information. Frequent "see" and "see also" references lead the user to related terms. In the author section the title is given under the name of the first author. Names of other authors are found in their alphabetical place with a "see" reference to the first author.

Other Material: Brief explanatory material is found in the preliminary pages.

Periodicals Scanned: A list is provided in each issue and cumulations.

AEROSPACE MEDICINE AND BIOLOGY see SCIENTIFIC AND TECHNICAL AEROSPACE REPORTS

ANIMAL DISEASE OCCURRENCE, 1980-
Commonwealth Agricultural Bureaux. Biannual.

Arrangement: The journal is divided into two sections: 1) abstracts, and 2) tables of data from the abstracts in section one. Material in the tables is tied to the source by the entry number of the abstract. The abstract section is arranged under twelve species headings.

Coverage: The introduction states that the journal contains "All recorded information (including that available in published and unpublished literature) on disease occurrence and the cost in animals, where it is of economic or public health importance. The range of species includes domesticated mammals, poultry, bees, and fish. Relevant information from journals, reports, congress proceedings, theses and bulletins of notifiable diseases have been included...."

Scope: International.

Locating Material: Entries are numbered consecutively through each volume. These numbers are referred to in the indexes and in the tables in section two.

Abstracts: The abstracts are in English. The titles are in English and in the language of the article. The language and summaries in other languages are indicated. The number of references is noted. The senior author's affiliation is given.

Indexes: An author and a subject index are found in each issue.

Other Material: Introductory and explanatory material is provided in four languages: English, French, German, and Italian. A user guide for the abstracts and the tables is provided. A "Retrospective analysis of the literature of animal diseases in Europe, January 1977 to July 1979" is found in Volume 1, no. 2.

Database: Entries available through CAB ABSTRACTS.

BRITISH MEDICINE, 1972-
    Pergamon Press. Monthly.

Arrangement: Arranged by form. That is, books are covered in the contents, then book reviews, nonbook materials, congresses, news and notes, and current periodicals. The book section is arranged by subject and then alphabetically by author and the current periodicals are alphabetical by title.

Coverage: Clinical medicine, surgery, biomedical sciences, nursing, pharmacy, psychology/psychiatry, health education, and social welfare are included among the subjects covered. A note in the preliminary pages states that the purpose of the journal is "...to provide a guide to new books and nonbook materials, pamphlets, official publications, brochures, and reports by research institutions and voluntary societies; to list the main contents of current issues of British medical periodicals and to note forthcoming international congresses."

Scope: British publications.

Locating Material: Entries in the books section are numbered consecutively through each issue. A logical approach would be by subject for books, and by journal title for periodicals.

Abstracts: Not really applicable. An outline of the contents of books is provided, authors and titles of articles in periodicals are given, and location and dates of forthcoming congresses are found.

Indexes: A combined author/title index to the book section is in each issue.

Other Material: Government publications, annual reports, and a section for general reading is found. Very brief explanatory notes are provided.

CALCIFIED TISSUE ABSTRACTS, 1969-
Cambridge Scientific Abstracts. Quarterly.

Arrangement: Arranged by subject.

Coverage: The contents include such subjects as bone composition, growth and tumors, cartilage, crystallization and crystal growth, diabetes, fracture, hormones, metabolism, osteopathies, plasmas, potassium, radioisotopes and radiation, sodium, teeth, tissue culture, trace elements, and vitamins.

Scope: International.

Locating Material: Entries are numbered consecutively through each volume. There are three elements to each number; the sequential abstract number, a code letter identifying the abstract journal, and the volume number.

Abstracts: Up to ten authors are given. Titles are in English and in the language of the article and the abstracts are in English. The introductory material states that "abstracts ... are usually 150-200 words in length, and outline the contents of the paper, the method used, the results obtained, and the conclusions drawn."

Indexes: An author index is included in each issue which cumulates annually.

Other Material: Explanatory material which includes a code for language abbreviations is found in each issue.

Periodicals Scanned: A list appears occasionally and is available on request.

Database: Entries are available through IRL LIFE SCIENCES COLLECTION from 1978. The updates are monthly and abstracts and keywords are provided.

CANCER THERAPY ABSTRACTS, 1960-1979.
Herner Information Services, Inc./National Cancer Institute. Ceased. Continues Cancer Chemotherapy Abstracts (1960-1964).

Arrangement: Arranged by subject.

Coverage: Clinical studies of neoplasms of all parts of the body, pharmacology, biochemistry, screening articles and reviews are found. Various therapies and treatments are covered.

Scope: International.

Locating Material: Entries are numbered consecutively from Volume 15, no. 1 (when the title changed; Cancer Chemotherapy Abstracts had a separate set of numbers). These numbers are referred to in the indexes.

Abstracts: Titles and abstracts are in English. An abbreviation standing for the language of the article is given in parentheses after the title. The senior author's affiliation is given. Abstracts are usually long and quite detailed and the number of references is given. "See also" references at the end of each section lead the user to related items.

Indexes: Author and subject indexes are found in each issue. These cumulate annually.

Other Material: A list of abbreviations used in the abstracts is provided. A concise but very useful description of the publication is furnished.

Periodicals Scanned: No list is provided but a note in the introductory material states that the abbreviations used are those of the "List of Journals Indexed in Index Medicus, List of Periodical Title Word Abbreviations."

CUMULATIVE INDEX OF HOSPITAL LITERATURE see HOSPITAL LITERATURE INDEX

CUMULATIVE INDEX TO NURSING AND ALLIED HEALTH LITERATURE, 1956-
The Seventh-Day Adventist Hospital Association. Five bimonthly

issues are cumulated in an annual hardbound edition (there is no issue for November/December but these entries are included in the cumulation). The following cumulations are available: Volume 1-5 in one volume, Volume 6-8 in one volume, Volume 9-11 in one volume and Volume 12-13 in one volume. Formerly: <u>Cumulative Index to Nursing Literature</u>.

<u>Arrangement</u>: Arranged by subject in the first section and by author or name in the second section. There is an appendix containing book reviews, films, pamphlets, etc.

<u>Coverage</u>: A broad selection of subject matter in medical and paramedical fields, especially nursing, its functions, technique in specific diseases, and under various circumstances, nursing in other countries, and topics allied to nursing and health, including health education, medical records, and physical and occupational therapy. Medical journals as well as popular journals are scanned for articles pertinent to the subject.

<u>Scope</u>: English language material.

<u>Locating Material</u>: Although there is no table of contents, the subjects covered are arranged alphabetically. An abundance of "see" and "see also" references guide the user to alternative terms. These references are included in the cumulative volumes even if that subject is not used in that volume. Thus the user is directed to other cumulative volumes or bimonthly issues. The author section is alphabetical. The "Subject Headings and Cross References" section leads the searcher to correct terms.

<u>Abstracts</u>: Not applicable. Entries in the subject section give the title of the article, author's name in parentheses and journal citation. In the author section, both authors' names are given if there are two, first author only if there are more than two, and titles and journal citations are found.

<u>Other Material</u>: A description with examples of how to use the journal is in each issue and the annual cumulation. A slide/tape explanation is available separately. There is a guide to Regional Medical Libraries in the annual cumulation.

<u>Periodicals Scanned</u>: A list is provided in each cumulated volume and the bimonthly issues. It gives abbreviations and full title, with complete address of the publisher in the annual volume. Also provided is an added list of full titles and abbreviations for ancillary journals scanned for articles useful to the nurse. An asterisk in the first list by a title indicates that only selected articles are indexed. An asterisk by a title in the second group means that articles from

that journal have been used and the volume number, beginning page and date are given.

CURRENT LITERATURE IN FAMILY PLANNING, 1972-
Planned Parenthood Federation of America, Inc. Monthly.

Arrangement: Arranged by subject within two parts: Part I, books, and Part II, articles.

Coverage: Religious and ethical views, sexual behavior, teenagers, sex information for children, various aspects of birth control and methods are among the subjects covered.

Scope: Mostly material from the United States, but publications from other countries will be reported if the content is of general interest.

Locating Material: The user searches by subject in his area of interest. The numbers of the entries are book call numbers of the Federation's library, the Katherine Dexter McCormick Library.

Abstracts: Concise but informative abstracts are provided. The titles and abstracts are in English. The number of pages, price and International Standard Book Number (ISBN) are included for books. The library will supply copies of articles and an order number is provided.

Other Material: An order form for articles is supplied. There is explanatory material in the preliminary pages.

CURRENT LITERATURE ON AGING, 1957-
National Council on Aging. Quarterly.

Arrangement: Arranged by subject, then alphabetically by author under each subject.

Coverage: Various aspects of treating the aging such as health and health care, rehabilitation of the handicapped, terminal care, social isolation and social participation, retirement planning, and mobility of the aged.

Scope: Principally U.S. journals are indexed and only English language material. Many U.S. government documents are included as well as books and journal articles.

Locating Material: Each entry is given a five-digit citation number. These began in Volume 15 with 72001. The first two digits refer to the year and the rest is a consecutive number which continues from that time. These are referred to in the indexes.

Abstracts: Usually very brief statements are given about the contents of an article. An indication that there are references is made if this is significant, but not how many.

Indexes: Annual author and subject indexes are provided.

Other Material: Officers of the Council and the library staff list is provided.

Periodicals Scanned: A list is in each issue of the journals scanned for that issue. Publishers' addresses are provided.

DENTAL ABSTRACTS, A SELECTION OF WORLD DENTAL LITERATURE, 1956-
American Dental Association. Monthly.

Arrangement: Arranged by subject.

Coverage: Selective coverage of dental literature including such subjects as anesthesiology, biochemistry, cavities, dentures, education, prevention, dental education, clinical dentistry, and dental practice.

Scope: International.

Locating Material: No numbering system is used. Index references are to page number.

Abstracts: Each abstract article is given a descriptive subject heading which is followed by the summary of findings in the article, followed by a full bibliographical citation. The location of the senior author is indicated.

Indexes: Annual name (author) index and subject index are provided.

Other Material: An article is found in each issue. Also in each issue are features called "News" and "News from NIH" and a list of new periodicals and new books.

Periodicals Scanned: A list appears annually which gives full title, abbreviations and the place of publication.

DEVELOPMENTAL DISABILITIES ABSTRACTS, VK V. 12-13, 1977-78.
U.S. Department of Health, Education and Welfare, Office of Human Development, Developmental Disabilities Office. Continues Mental Retardation Abstracts, 1964-1976. Ceased.

Arrangement: A loose subject arrangement is employed with subheadings under broad topics.

Coverage: Medical, development, treatment, training and programmatic aspects of mental retardation are covered. Such specific topics as convulsive disorders, psychodiagnosis, therapy, etc., will be found (see Table of Contents). Books, symposia, and proceedings are included.

Scope: International.

Locating Material: Entries are numbered consecutively through each volume. These numbers are referred to in the indexes.

Abstracts: The titles are in English. The abstracts are also in English and are usually detailed, giving a good idea of the content of an article. The number of references and the source of the abstract is given. All authors are listed and the senior author's affiliation is given.

Indexes: Author and subject indexes are in each issue. These cumulate annually.

Other Material: Some issues include a bibliography on a specific related subject. These are found in the preliminary pages of the issues which carry them.

Periodicals Scanned: A list of journals and books is provided in Volume 13, number 4 (1978), the final issue.

DRUG ABUSE BIBLIOGRAPHY, 1971-
Whitson Publishing Company. Annual. (Annual Supplements to Drugs of Addiction and Nonaddiction; Their Use and Abuse: A Comprehensive Bibliography, 1960-1969.)

Arrangement: The bibliography is divided into four sections: 1) books and government publications, 2) title index to periodical literature, 3) subject index to periodical literature, 4) author index. The first section is alphabetical by author, the second, alphabetical by title, the third, alphabetical by title under each subject heading, and finally alphabetical by author.

Coverage: Drugs or controlled substances by name are covered, detection, control, consequences of abuse of various kinds of drugs, social problems and social groups are found and drug abuse in various countries is given.

Scope: International.

Locating Material: No numbering system is used. One searches by subject or author and title. The author index refers to page numbers.

Abstracts: Not applicable. Article titles are given in English. Only the senior author is provided. Publication information is given for monographs.

Indexes: Author index is found in each volume.

Other Material: A brief description of the index is found in the preface and a list of the material searched to compile each volume. A list of subject headings is provided.

Periodicals Scanned: A list is included in each volume.

DSH ABSTRACTS, 1960-
Deafness, Speech and Hearing Publications, Inc. Quarterly.

Arrangement: Arranged by subject. Alphabetical by author within each subject section.

Coverage: Hearing and speech and related disorders are covered. Major headings are divided into more specific areas of interest as: anatomy and physiology; instruments and procedures, diagnosis and appraisal, aetiology and pathology, medical and surgical treatment, speech-reading and manual communication, speech and language development; voice, aphasia, delayed speech, stuttering, etc. Books as well as journal articles are included.

Scope: International.

Locating Material: Entries are numbered consecutively through each volume. These numbers are referred to in the contents (with page numbers for the beginning) and in the index.

Abstracts: Titles are in the language of the article with an English translation. Languages of articles other than English are named. Abstracts are in English. The first author's address and the source of the abstract are given. Abstracts range in length from quite long and detailed to very brief.

Indexes: An author index is found in each issue (which cumulates annually), and an annual subject index is in the final issue.

Other Material: A list of abstractors is found in each issue. A list of abbreviations employed is in the annual subject index.

Periodicals Scanned: A list is in the final issue of each volume.

EXCERPTA MEDICA, 1947-
Excerpta Medica Foundation. Published in sections (each having a code number). Frequency for the different sections varies as does the beginning date. Nineteen forty-seven represents the first time the publication appeared. Each section covers a particular specialty and some abstracts appear in more than one if they are relevant. Titles of sections have varied over the years, usually because of splitting a section or enlarging the title. Sections presently being published are listed below (section number follows the title); "title varies" notes have been omitted:

Adverse Reactions Titles (38)
Anatomy, Anthropology, Embryology and Histology (1)
Anesthesiology (24)
Arthritis and Rheumatism (31)
Biophysics, Bio-engineering and Medical Instrumentation (27)
Cancer (16)
Cardiovascular Diseases and Cardiovascular Surgery (18)
Chest Diseases, Thoracic Surgery and Tuberculosis (15)
Clinical Biochemistry (29)
Dermatology and Venereology (13)
Developmental Biology and Teratology (21)
Drug Dependence (40)
Drug Literature Index (37)
Endocrinology (3)
Environmental Health and Pollution Control (46)
Epilepsy (50)
Forensic Science (49)
Gastroenterology (48)
General Pathology and Pathological Anatomy (5)
Gerontology and Geriatrics (20)
Health Economics and Hospital Management (36)
Hematology (25)
Human Genetics (22)
Immunology, Serology and Transplantation (26)
Internal Medicine (6)
Leprosy and Related Subjects (51)
Microbiology (4)
Neurology and Neurosurgery (8)
Nuclear Medicine (23)
Obstetrics and Gynecology (10)
Occupational Health and Industrial Medicine (35)
Ophthalmology (12)
Orthopedic Surgery (33)
Otorhinolaryngology (11)
Pediatrics and Pediatric Surgery (7)
Pharmacology and Toxicology (30)
Physiology (2)
Plastic Surgery (34)

Psychiatry (32)
Public Health, Social Medicine and Hygiene (17)
Radiology (14)
Rehabilitation and Physical Medicine (19)
Surgery (9)
Urology and Nephrology (18)
Virology (47)

Arrangement: Arranged by broad subjects according to a detailed and specifically designed classification. An outline of the classification appears on the inside front cover of each issue.

Coverage: Covers original articles appearing in the world's leading biomedical journals. Areas excluded are dentistry, nursing, and veterinary medicine. Even though veterinary medicine is excluded, many veterinary journals are screened for articles on comparative pathology and epidemiology. Also excluded are psychology and other such paramedical fields as podiatry, optometry, and chiropractic. The sections on Health Economics and Environmental Health go considerably beyond the scope of biomedicine. The former covers the financial aspects of health care, public health policies, and hospital management. The latter covers all aspects (biological, chemical, economic, legal, medical, sociological, and technological) of air, water, and soil pollution, noise hindrance, and radioactivity.

Scope: International.

Locating Material: Items are numbered consecutively through the volume. Within each subject category, the abstracts are numbered in sequence of their accession to the Excerpta Medica database. These numbers are used in the indexes.

Abstracts: Titles are in English and in the language of the article. Abstracts are in English. Up to four authors are given and the senior author's address is provided. Abstracts are usually detailed.

Indexes: Author and subject indexes are in each issue. These cumulate annually. The primary author and up to three joint authors are listed in the author index (if there are more than four, only the first three are named). The subject index terms are divided into primary and secondary terms, that is, first by important concept, under which more detailed aspects of the concept are found. For example: under the broad subject, "Open Heart Surgery," will be found such subheadings as "acute kidney failure," "anesthesia," "bacterial endocarditis," and other papers relating to the broader subject. Also, a paper may appear under more than one broad concept subject heading. Primary terms are from a thesaurus called MALIMET (Master List of Medical Indexing Terms) which was designed for computerized indexing of the subjects covered.

Other Material: A list of offices and distributors of Excerpta Medica, with their addresses, is in each issue. Published separately is a Guide to the Use of Excerpta Medica Abstract Journals: A List of 4,000 Biomedical Terms Most Commonly Used in Search Formulation of Secondary Publications. Also published separately is Adverse Reactions Titles and Drug Literature Index.

Database: Entries can be retrieved from EMBASE from 1974. The update is monthly and abstracts and keywords are provided.

GERONTOLOGICAL ABSTRACTS, 1976-
University Information Services, Inc. Six times a year in two volumes.

Arrangement: Arranged by broad subjects subdivided into narrower topics.

Coverage: The broad subjects are biological, clinical, and social aspects. The subdivisions include cell biology, endocrinology, physiology, genetics, nutrition, heart, blood vessels, nervous system, respiratory and digestive systems, psychology and psychiatry, immunology, learning, memory and intelligence, theories of aging, retirement, sex, and the quality of life.

Scope: International.

Locating Material: Abstracts are numbered consecutively (with a seven-digit number) through the volume.

Abstracts: Key terms are provided at the beginning of each entry. The abstracts are in English and the source journal is named at the end of the abstract.

HOSPITAL ABSTRACTS, 1961-
Her Majesty's Stationery Office (prepared by the Department of Health and Social Security). Monthly.

Arrangement: Arranged by subject. There are broad headings with subtopics.

Coverage: Hospitals, planning and design, equipment, staff, organization and administration, finance and accounting, supplies and services, dietetics, hygiene, safety, the patient, and special hospitals are included (see Contents). The whole field of hospitals and their administration, except strictly medical material, is found. Monographic material is included.

Scope: International.

Locating Material: The entries are numbered consecutively through each volume. These numbers are referred to in the indexes.

Abstracts: Abstracts are in English. Titles are in the language of the article and in English. The bibliographic citation includes the place of publication of the journal. Abstracts are usually fairly detailed giving a good idea of the content of the article. Frequent "see" and "see also" references at the beginning of each subject section refer the user to related entries.

Indexes: Author and subject indexes are found in each issue. These cumulate annually.

Other Material: Brief explanatory information is found on the verso of the front cover.

Periodicals Scanned: A list is in each issue but includes only publications referred to in that issue. When known, the price of a single issue is given. The list gives the full journal title and its address.

HOSPITAL LITERATURE INDEX, 1945-
American Hospital Association. Semiannual with five-year cumulation, 1945-1961; Quarterly with annual and five-year cumulations, 1962- . The five-year volume entitled, Cumulative Index of Hospital Literature was last published in 1979 and covered the years 1975-1977.

Arrangement: Arranged in two sections, subject and author. The subjects are alphabetical and are rather broad with subdivisions for narrower topics.

Coverage: Administration of hospitals and other health-care institutions, their planning and financing are covered. Such topics as cost accounting, automation, human relations, paramedical personnel, statistics, etc. are included. Special problems such as intensive care, psychiatric care, surgical nursing, quality assurance and risk management are also found.

Scope: Only English language material is covered, but it may be from any country of the world.

Locating Material: As the subject headings are alphabetical and have subdivisions for more specific topics, the reader searches for the proper heading. Entries under the subject headings are alphabetical by title. "See" and "see also" references guide the user to the alternative terms. Author sections are alphabetical by last name.

Abstracts: Not applicable. Entries include author, title, and journal citation.

Other Material: A description of how to use the index is in the preliminary pages of quarterly issues and in annual and five-year cumulations. A list of recent acquisitions of the Library of the American Hospital Association including books, monographs, and journals is found as the last section in the quarterly issues and in the annual cumulation.

Periodicals Scanned: A list is in the quarterly issues, annual and five-year cumulations.

INDEX CATALOGUE OF MEDICAL AND VETERINARY ZOOLOGY, 1932-
Oryx Press. Formerly U.S. Department of Agriculture, Agriculture Research Service (U.S. Government Printing Office). Irregular. A revision and continuation of Index Catalogue of Medical and Veterinary Zoology Authors, published 1902-1912 as Bureau of Animal Industry Bulletin 39. Issued in a series of publications. Initially covered authors' names A-Z, and subsequent numbers were in the nature of supplements (revisions were published until 1952). In 1953 a series of supplements was begun in order to publish the backlog. This was completed with supplement six (1956). Since then supplements covering authors A-Z are issued on an annual or biennial basis. From supplement fifteen, Parasite-Subject Catalogues have been issued.

Arrangement: Part 1 is arranged alphabetically by the author's name. Complete bibliographical information necessary to locate an article is found here. Personal authors (individuals) and corporate authors (companies, universities, laboratories, etc.) are interfiled. At the end of each entry there is an abbreviation enclosed in brackets. This is an indication of a library which owns the journal listed. A key to these abbreviations and to serial title abbreviations is found in the front of Part 1, but unfortunately the list in any one issue is not always complete, so one must occasionally consult more than one supplement. Sometimes an entry will have a superior numeral. These are used to distinguish between two authors of the same name when the surname is unknown. In parts 2-7, parasite entries are alphabetical by genera, parasite diseases, higher taxa, then by species within genera. Host entries are by genera, species within genera and common names. Subject headings and treatments used for parasitic diseases are alphabetical.

Coverage: A wide range of information relating to the subject is included. Parasites listed are Protozoa, Trematoda and Cestoda, Nematoda and Acanthocephala, Arthropoda, miscellaneous phyla. Treatments and hosts are also covered.

Scope: International.

Locating Material: The searcher may take an author approach if the author's name is known, but this is rare. Primarily one works from the detailed indexes back to Part 1, the Author Catalog. If more than one article is listed under an author's name, a year plus letter code tells the searcher which one refers to his subject. Entries in Parts 2-5 of the Subject Catalog are in two double columns (four altogether). The author's name is given to the right of the subject entry. Information in the subheadings in the left column (indented under the subject heading) gives details bearing on classification, hosts, treatment, etc. Subheadings in the right half of the column (under the author entries) give geographic distribution. The subjects in parts 2-5 are parasites, i.e., Part 2, Protozoa; Part 3, Trematoda and Cestoda; Part 4, Nematoda and Acanthocephala; Part 5, Arthropoda and Miscellaneous Phyla.

Entries in Parts 1, 6, and 7 are in two columns. Part 6 is the Subject Headings and Treatment Section. Each entry consists of an entry term or subject, author's name and the necessary codes to lead one to the proper bibliographic information in Part 1. Subheadings reflect the information given in the article about the subject. Part 7 is the Host Catalog and the hosts recorded are those that pertain directly to the author's own work. Scientific host names are used unless the author gave only common names, in which case the host names are given as found in the publications. Each entry consists of the name of the parasite or parasitic disease, the author's name and the subheading which gives additional information.

Abstracts: Not applicable. Entries are in English and in the language of the article. Illustrative material is indicated. Numerous "see" and "see also" references guide the searcher to other entries in the Author Catalogue. Taxonomic articles are indexed in depth; new taxa, new combinations, synonyms, keys and taxonomic revisions are indicated. Experimental articles are analyzed under subject headings and treatments as well as names of parasites and parasitic diseases. Host records are indexed in detail by scientific names of host animals.

Other Material: A description of how to use the catalogue is included in the preliminary pages of each part. Special publications are issued from time to time covering specific subjects.

INDEX MEDICUS, 1960-
> National Library of Medicine (U.S. Government Printing Office). Monthly with annual cumulation called Cumulated Index Medicus (the cumulated volume was previously issued by the American Medical Association). Supersedes Current List of Medical Literature 1941-1959. For earlier material one should consult: Index Catalogue of the Library of the Surgeon General's Office, 1880-1955; Index Medicus, A Monthly Classified Record of the Current Medical Literature of the World, issued by the Carnegie Institution, Volume 1-21, 1879-1899; Ser. 2, Volume 1-18, 1903-1920; Ser. 3, Volume 1-6, no. 5, 1921-1927. This publication was suspended between 1899 and 1902. During this time, Bibliographia Medica was issued by the Institut de Bibliographie of Paris. The American Medical Association also issued a Quarterly Cumulative Index Medicus from 1927-1956 which was a merging of the Quarterly Cumulative Index and the old Index Medicus. The description below pertains only to Index Medicus, 1960-

Arrangement: Entries are arranged under specific subject headings (a list of subject headings is provided) and are found under as many headings as necessary to describe their content adequately. An author section provides a means of access to material by author's name. A Bibliography of Medical Reviews is provided monthly with an author and subject section. This contains "well documented surveys of recent biomedical literature."

Coverage: An extremely broad range of subject material related to medical and health sciences is included. Only periodical literature is covered. Letters, editorial biographies and obituaries are found if the content merits. Abstracts of articles are not indexed nor are proceedings of conferences and symposia, etc., unless they are published in periodicals. The list of subject headings is included as Part 2 of the January issue and is found annually in Cumulated Index Medicus. This provides the user with a comprehensive list of descriptors. There is also a Medical Subject Headings Annotated Alphabetic List available from NTIS.

Scope: International.

Locating Material: The searcher should consult the list of subject headings to select the term which most closely approximates his concept. In each issue and in the annual cumulation the entries are arranged in alphabetical order. The searcher is guided to alternative terms by means of a letter-number code. In Part 2 of the January issue of each volume and in the annual cumulation are two lists: an alphabetical list of the subject headings arrangment; and a categorized list of the same subject headings arranged by this code. When a searcher finds a particular subject he is guided to other related terms by means of this code. One looks under the code in the list and discovers the precise, alternative term and is thus guided to other references. Authors are listed alphabetically in a separate section.

Abstracts: Not applicable. Titles are in English. The entries include author, title, journal citation, an abbreviation indicating the language of the article if other than English (a key is provided in the Introduction). If abstracts are to be found in English in the source periodical, this is noted.

Other Material: An extensive introduction in each issue and the annual cumulation gives a good description of how to use the index and what to expect from it. A list of abbreviations for foreign languages is found in the Introduction. A list of subject headings formerly used with the term presently employed and a comprehensive list of the subject headings with cross references are given. Medical subject headings also appear as Part 2 of the first issue. A Bibliography of Medical Reviews is provided. A description of the access to articles, a list of Regional Medical Libraries and MEDLARS (Medical Literature Analysis and Retrieval System) Centers is provided. Following is a list of indexes and bibliographies extracted from Index Medicus by means of MEDLARS. In general, the format is the same and the citations found in these publications will also be found in Index Medicus except for Index to Dental Literature and International Nursing Index. These two contain journal citations not indexed for Index Medicus. Most of this material is printed and distributed by nonprofit professional organizations (see list on the inside back cover of Index Medicus).

Anesthesiology Bibliography
Annual Bibliography of Orthopedic Surgery.
Bibliography of Acute Diarrhoeal Diseases
Bibliography of Podiatric Medicine and Surgery
Bibliography of Surgery of the Hand
Bibliography on Medical Education
Cranio-Facial-Cleft Palate Bibliography
Current Bibliography of Plastic and Reconstructive Surgery
Current Citations on Strabismus, Amblyopia, and Other Diseases of Ocular Mobility
Family Medicine Literature Index
Hospital Literature Index
Index of Audiovisual Serials in the Health Sciences
Index of Rheumatology
Index to Dental Literature
International Nursing Index
Neurosurgical Biblio-Index
Physical Fitness/Sports Medicine
Psychopharmacology Bibliography
Quarterly Bibliography of Major Tropical Diseases
Recurring Bibliography of Hypertension
Recurring Bibliography on Education in the Allied Health Professions

(Some of these titles are covered elsewhere in this book.) The National Library of Medicine Current Catalog, published quarterly and cumulated annually and quinquennially is also useful for medical material. This contains citations to all publications cataloged by the Library during the period covered (except for titles published before 1801).

Periodicals Scanned: A list is in the annual cumulation and in Part 1 of the first issue of each volume. It gives the abbreviated title as used in the entries, the full title and the place of publication.

Database: Citations are available through MEDLINE since 1977, with 1966-76 available in backfiles. Abstracts are provided but not keywords. The update is monthly. (More accurately the database is MEDLARS and the retrieval system is MEDLINE.)

INDEX VETERINARIUS, 1933-
Commonwealth Agricultural Bureaux, prepared by Commonwealth Bureau of Animal Health. Quarterly, 1933-1971, Monthly from 1972.

Arrangement: A separate subject and author index, each in alphabetical order is found.

Coverage: Serial publications, books, annual reports, monographs, theses, and other nonserial publications are covered. All major items of veterinary literature are listed in Index Veterinarius and abstracted in The Veterinary Bulletin. "In addition the following categories of titles are listed in Index Veterinarius only:

1. Marginal fields of veterinary medicine, such as history and jurisprudence.

2. Minor contributions, such as case reports, brief reviews, obituaries, correspondence, abstracts of conference papers, notes and news items.

3. Literature of the extension type, written for students and farmers.

4. Repetitive literature, such as papers duplicating a previously published report, and small-scale trials of a drug or a method which has already become established.

5. Related fields of other biological sciences, such as pure microbiology, general parasitology, general pathology, human medicine, and biochemistry.

6. Literature of purely local interest, such as that describing the disease or parasite situation within a small territory.

7. Publications received too late (a year or more after publication) for abstracting.

8. Chapters in books."

Scope: International.

Locating Material: An alphabetical arrangement is used; see above.

Abstracts: Not applicable. A complete bibliographical citation is given. The original language of an article is shown under the author entry. Under the subject entries, an English translation is given. Exceptions are the Cyrillic, Arabic, Japanese, and Chinese scripts, for which only an English translation is given in the author and subject entries. In addition, the language of the article and of any available summaries is noted. Each title is listed under one or more subject headings in the subject section of the index and also in the author section. When a paper has more than one author, cross references are provided for joint authors. Author's addresses are found in the author section. Subject headings are derived from "Veterinary Multilingual Thesaurus" published in 1979.

Other Material: Introductory material includes a statement on the method of selection of titles, the system of entry of titles, subject headings, languages used, and provides a list of abbreviations for languages.

Periodicals Scanned: A list appears in the January 1983 issue of Veterinary Bulletin and copies are available separately from the Bureau.

Database: Entries are available through CAB ABSTRACTS from 1972 and are updated monthly. Abstracts and keywords are provided. The other abstract journals of the Commonwealth Agricultural Bureaux are also included.

INTERNATIONAL NURSING INDEX, 1966-
American Journal of Nursing Company. Quarterly. (The first three issues of each volume are paperbound and the fourth is a clothbound cumulation.)

Arrangement: Arranged in two sections: 1) Subject and 2) Name. In the subject section occasionally a main topic will be subdivided into narrow aspects for better coverage. Each article is indexed under three or more subjects.

Coverage: A comprehensive indexing of nursing literature that includes all facts of the field is supplied here. Various categories such as surgical nursing, public health nursing, geriatric nursing, and school nursing are found. Problems, techniques, and research in these various categories as well as specific diseases or health problems are covered.

Scope: International.

Locating Material: No numbering system is used. The searcher uses the subject approach and as articles are indexed under at least three subjects the coverage is thorough. If an author is known, this access may be used.

Abstracts: Not applicable. Entries include senior author, title in English and an indication of the original language for articles in another language; and a complete bibliographic citation is given. This information appears each time the title is included.

Indexes: A Thesaurus appears in the cumulative volume which guides the reader to proper subject terms used as headings. There are cross references from unused terms to those which are employed.

Other Material: A detailed description and explanation of use appears in each issue. A list of publications of nursing organizations and agencies is included in each issue and in the cumulative volume. A list of nursing books arranged alphabetically by the name of the country (which represents those reviewed by the publisher) is also included.

Database: Entries are available through MEDLINE (MEDLARS) from 1966. The update is monthly.

INTERNATIONAL PHARMACEUTICAL ABSTRACTS, 1964-
American Society of Hospital Pharmacists   Semimonthly.

Arrangement: Divided into 24 sections (25 with the subject index) each of which covers a different broad topic.

Coverage: Among the topics covered are pharmaceutical technology, institutional pharmacy practice, adverse drug reactions, investigational drugs, drug evaluations and interactions, biopharmaceutics and pharmaceutics, drug stability, pharmacology, preliminary drug testing, pharmaceutical chemistry, drug analysis, drug metabolism and body distribution, microbiology, pharmacognosy, methodology, environmental toxicity, legislation, laws and regulations, history, sociology, economics and ethics, pharmaceutical education, pharmacy practice, and information processing and literature.

Scope: International.

Locating Material: Each entry has a six-digit number with a space after the first two digits (these represent the volume number). The numbers run consecutively through the volume and are referenced in the author and subject indexes.

Abstracts: Titles and abstracts are in English. If the article is in another language that language is noted in italics. Summaries in other languages are indicated by an abbreviation. There are some title-only entries so that articles not pertinent enough for an abstract but too important for omission can be included. The number of references is given.

Other Material: A thorough description of how to use IPA is included with a description of what kinds of abstracts to expect under the subjects. A list of language abbreviations is provided and a price list for a reprint service is given.

Periodicals Scanned: The first issue of each volume contains a list of journals covered in the previous year. The full journal name, the international CODEN designation and the abbreviation used in the abstracts are provided.

Database: Entries are available through INTERNATIONAL PHARMACEUTICAL ABSTRACTS (IPA) from 1970 and are updated monthly. Abstracts and keywords are provided. International Pharmaceutical Abstracts can also be accessed through TOXLINE.

LEUKEMIA ABSTRACTS, 1953-
Lenore Schwartz Leukemia Research Foundation, Research Information, John Crerar Library. Monthly.

Arrangement: Arranged alphabetically by the author of the article.

Coverage: Leukemia, related medical and scientific problems, and allied diseases are included. Effects of treatment, comparison of patients and current research are covered.

Scope: International.

Locating Material: Entries are numbered consecutively through each volume, but no index is provided.

Abstracts: The abstracts are in English. The titles are in English but if the article was originally in another language, they are in parentheses and the name of the language is given. The abstracts are brief and concise, but give good information concerning the content of an article. Occasionally none appears. Affiliation of the senior author is given and usually the number of references.

MEDITSINSKII REFERATIVNYI ZHURNAL, 1960-
    Medgiz. Monthly. (Issued in 22 sections.)

Arrangement: Arranged by broad topics subdivided into narrow
    subjects.

Coverage: All aspects of medicine and disease including surgery,
    anesthesiology, hygiene, obstetrics, endocrinology, hema-
    tology, medical genetics, and laboratory investigation are
    covered. A list of the sections' contents is provided in each
    issue.

Scope: International.

Locating Material: Entries are numbered consecutively through each
    volume. These numbers are referred to in the indexes.

Abstracts: The abstracts are usually quite long and they are in
    Russian. The titles of articles in the Cyrillic alphabet are
    repeated in English. Articles in the Latin alphabet languages
    are given in the original. For titles in other non-Latin languages,
    the language and the country of origin are noted.

Indexes: There are author and subject indexes for each section
    which cumulate annually.

MENTAL HEALTH BOOK REVIEW INDEX, 1956-1972.
    Council on Research in Bibliography, Inc. Ceased.

Arrangement: Arranged alphabetically by author or corporate entry.

Coverage: Reviews of monographic literature in the field of mental
    health, psychology, psychiatry, and behavioral sciences.

Scope: Reviews appearing in English language periodicals, some
    originating outside the United States are included.

Locating Material: Entries are numbered consecutively beginning
    with the first issue. It frequently happens that an entry will
    fall in the correct alphabetical order but the number will be
    out of sequence. This means that this book has been pre-
    viously listed and now has additional reviews (three or more)
    and is listed with the original number. The subsequent num-
    bers under the entry refer to the issue number in which the
    publication was originally listed (marked *) and in which the
    listing was repeated (marked R).

Abstracts: Not applicable. Entries contain full bibliographic in-
    formation, journal citation for review and the reviewer's
    name. The number of pages, price and extensive title page

information is given. This is provided the first time the publication is listed. Issue numbers under the original entry indicate other reviews and bibliographic data.

Indexes: Cumulative author-title index for issues 1-12 (1956-1967) was published in 1969.

Other Material: A description of how to use the index is provided in the preliminary pages of each issue and in the cumulative index. Some issues contain editorials. A list of these is provided in Number 13 (1968) covering the years 1959-1967.

Periodicals Scanned: A list is in each issue.

MENTAL RETARDATION ABSTRACTS see DEVELOPMENTAL DISABILITIES ABSTRACTS

MONTHLY BIBLIOGRAPHY OF MEDICAL REVIEWS see INDEX MEDICUS

MULTIPLE SCLEROSIS INDICATIVE ABSTRACTS, 1974-
Federation of American Societies of Experimental Biology. Bimonthly.

Arrangement: Arranged by subject.

Coverage: Etiology, diagnostic methods, pathogenesis, epidemiology and genetic studies, prevention, therapy, and patient management, general review articles, experimental allergic encephalomyelitis and viral models, relevant research on other demyelinating diseases, chemistry and immunology of myelin, biology of glial cells, and other basic research are listed as the subject headings covered.

Scope: International.

Locating Material: Entries are numbered consecutively.

Abstracts: Titles and abstracts are in English. The source of the abstract is given and addresses for ordering reprints are provided. A list of keywords is found at the end of each abstract.

Indexes: An author index is found in each issue.

Other Material: Very brief explanatory notes are given.

NATIONAL LIBRARY OF MEDICINE CURRENT CATALOG see INDEX MEDICUS

NEUROSCIENCES ABSTRACTS, 1983-
Cambridge Scientific Abstracts. Monthly.

Arrangement: Arranged by broad subjects subdivided into more narrow categories.

Coverage: Neuroanatomy, growth and development, aging, degeneration and repair, neurophysiology, neurochemistry, neuroendocrinology, neuropharmacology, neurotoxicology, immunology, genetics, experimental pathology, neural correlates of behavior and sleep and the broad subjects listed in the Contents.

Scope: International.

Locating Material: Entries are numbered consecutively through each volume. These numbers contain three elements: the sequential number, a code letter which identifies the abstract journal, and finally the volume number.

Abstracts: Titles are in English and in the language of the article. Up to ten authors' names are given and the senior author's affiliation is provided. Introductory material states that "abstracts ... are usually 150 words in length and outline the contents of the paper, the methods used, the results obtained and the conclusions drawn."

Indexes: Author and subject indexes are found in each issue which cumulate annually.

Other Material: Explanatory material which includes a code for language abbreviations is found in the preliminary pages.

Periodicals Scanned: A list of source journals is available on request.

Database: Entries are available through IRL LIFE SCIENCE COLLECTION from 1983. Updates are monthly and abstracts and keywords are provided.

ORAL RESEARCH ABSTRACTS, 1966-
American Dental Association. Monthly (two issues in April).

Arrangement: Arranged by subject.

Coverage: A note in the preliminary pages states that this journal "...provides abstracts of original articles from the dental and nondental literature which have a bearing on dentistry and its related arts and sciences." Among the subjects listed in the contents are: anatomy, anesthesiology, biochemistry, caries, forensic dentistry, operative dentistry, pathology and oral medicine, periodontics, pharmacology, public health, and radiology and radiography. Patents are also included.

Scope: International.

Locating Material: Entries are numbered consecutively through each volume.

Abstracts: The titles and abstracts are in English. If the article is in a language other than English, that language and summaries in other languages are indicated. Abstractors' names are provided and the senior author's affiliation is usually given.

Indexes: Each issue contains an author index. Annual cumulative author and subject indexes are provided.

Other Material: A list of abbreviations utilized in the publication is supplied.

PHYSICAL EDUCATION INDEX, 1978-
BenOak Publishing Co. Quarterly with the fourth issue a hardcover cumulation.

Arrangement: Arranged by subject; book review section is alphabetical by the author of the book.

Coverage: The preface states that comprehensive coverage is given to dance, health, physical education, physical therapy, recreation, sports, and sports medicine, and to specific subject areas such as administration, biomechanics-kinesiology, coaching, curriculum, facilities, history, measurement evaluation, motor learning, perception, philosophy, physical fitness, research, sport psychology, sport sociology, teaching methods, training, and sports activities. Book reviews are included in a separate section.

Scope: International.

Locating Material: Searcher should use a subject approach. No numbering system is used.

204 / Abstracts and Indexes

Abstracts: Not applicable. Entries include complete bibliographic information but only the senior author is named. An article is included under as many subjects as necessary to give it complete coverage.

Other Material: Explanatory material is found in the preface with an outline of criteria for what is and is not included. Lists of abbreviations are also found.

Periodicals Scanned: A list is found in the bound volume.

PHYSICAL FITNESS/SPORTS MEDICINE, 1978-
President's Council on Physical Fitness and Sports. For sale by the Superintendent of Documents. Quarterly.

Arrangement: Arranged by subject.

Coverage: Included are such subjects as energy, metabolism, vitamins, exercise, endurance, muscles and muscle contractions, parts of the body and the effect of exercise on them, exertion, protective clothing and devices, various kinds of sports and athletic injuries. The citations are retrieved from the MEDLARS database of the National Library of Medicine. Papers presented at some congresses are also found.

Scope: International, but only those foreign language publications that provide an English summary are included.

Locating Material: No numbering system is used. The entries are in alphabetical sequence under each subject according to the name of the journal in which they appear.

Abstracts: Not applicable. Titles are in English. A bracketed translation is provided if the article is in another language and a note is made of an English summary. An abbreviation is given which indicates the language.

Indexes: An author index is provided in each issue.

Database: As the entries are derived from MEDLARS, they are available through MEDLINE from the beginning of the index.

PSYCHOLOGICAL ABSTRACTS, 1927-
American Psychological Association, Inc. Monthly. From 1971, two volumes per year with six issues per volume. Preceded by Psychological Index, 1894-1936.

Arrangement: Arranged by subject: broad classification of major interest areas of psychology which are subdivided into narrower topics. Within each issue, abstracts are alphabetical by author in each subsection.

Coverage: General psychology, psychometrics, experimental psychology (both human and animal) developmental psychology, social issues, physical and psychological disorders, educational psychology and applied psychology. The subtopics include parapsychology, perception and motor processes, learning and motivation, psychosexual behavior and sex roles, health care services and human factor engineering.

Scope: International, though U.S. journals are in the majority.

Locating Material: Abstracts are numbered consecutively within each volume. The indexes refer the searcher to these numbers.

Abstracts: Abstracts are in English. Titles are in English and usually in the language of the article and available summaries in other languages are noted. The abstracts are concise and nonevaluative but occasionally none will be provided. Books and chapters are annotated. There are no abstracts for citations from other abstract publications. Bibliographic information is provided in full for all types of publications. The address of the first author of journal articles is provided. The number of references is included if there are fifteen or more (sometimes given as number of pages--e.g., $3\frac{1}{2}$ p. ref.).

Indexes: Author and brief subject index is found in each issue. The subject index consists of terms drawn from "Thesaurus of Psychological Index Terms." A cumulative author and subject index for each volume gives the heading terms, descriptive phrases, and abbreviations indicating the language of the article if other than English. Cumulative Subject and Author Indexes for 1927-1959, 1960-1963, 1964-1968, 1969-1971 are available.

Other Material: An overview of operations and description of how to use the journal and auxiliary services and a list of abbreviations used in the abstracts are provided. A guide describing use of the journal and including a list of abbreviations is also found. A list of abstractors is in the volume cumulative issues.

Periodicals Scanned: A list is in each volume cumulative index issue.

Database: Entries are available through PSYCHOLOGICAL ABSTRACTS (PSYCHABS) from 1967. Updates are monthly and keywords and abstracts are provided.

206 / Abstracts and Indexes

PSYCHOPHARMACOLOGY ABSTRACTS, 1961-1983.
National Clearinghouse for Mental Health Information of the National Institute of Mental Health (U.S. Government Printing Office). Ceased.

Arrangement: Arranged by subject. The material is divided into seventeen categories.

Coverage: Subjects cover preclinical psychopharmacology including synthesis, isolation, development, mechanism of action, toxicology, etc. and clinical psychopharmacology which includes drug trials in various disorders, mechanism of action, toxicology, etc. (See Table of Contents for a complete list.)

Scope: International.

Locating Material: Each entry is assigned an identifying number. In the indexes this number is given additional numbers in parenthesis which mean the issue number and the subject category.

Abstracts: Titles and abstracts are in English. Abstracts are usually concise, but sometimes quite detailed. All authors are given and the senior author's affiliation is provided. The place of publication of the journal, the number of references and the abstract source are provided. The format in this journal has changed over the years, but these essential elements have remained more or less constant.

Indexes: Author and subject (Keyword) indexes are in each issue. These cumulate annually.

Periodicals Scanned: A list is provided upon request.

PUBLIC HEALTH ENGINEERING ABSTRACTS, 1921-1967.
U.S. Department of Health, Education and Welfare (U.S. Government Printing Office). Ceased.

Arrangement: Arranged by subject.

Coverage: Atmospheric pollution, solid wastes, foods, health practices, radiological safety, water pollution, etc. are among the subjects covered (see Contents).

Scope: International.

Locating Material: Entries are numbered consecutively through each volume. These numbers are referred to in the indexes.

Abstracts: Titles and abstracts are in English but abstracts are not always provided. Author's address, number of references and illustrative material, source of abstract, and a list of index terms are given.

Indexes: There are annual author and subject indexes.

REHABILITATION LITERATURE, 1940-
The National Easter Seal Society. Bimonthly.

Arrangement: This journal contains articles, a Review of the Month, books reviewed and abstracts of the literature (see Table of Contents). The book review section is alphabetical by title. The abstract section is arranged alphabetically by subject.

Coverage: Kinds of disorders and methods of treatment are covered, such as autism and aphasia, behavior modification, chronic diseases, various kinds of therapy, and medical treatment of disorders.

Scope: U.S., British, and Canadian journals are indexed. Monographs are included in the abstract section.

Locating Material: Book reviews and abstracts are numbered consecutively through each volume (book reviews are first).

Abstracts: Usually quite lengthy and detailed. Senior authors and addresses are given. "See references" guide the user to related or alternative items.

Indexes: There is an author index in each issue which cumulates annually.

Other Material: Article of the Month, special articles, events, commentary and a Review of the Month are in each issue (see Table of Contents).

REVIEW OF APPLIED ENTOMOLOGY, SER. B., MEDICAL AND VETERINARY see REVIEW OF APPLIED ENTOMOLOGY

REVIEW OF MEDICAL AND VETERINARY MYCOLOGY COMPILED FROM THE WORLD LITERATURE ON MYCOSES OF MAN AND ANIMALS, 1943-
Commonwealth Agricultural Bureaux, prepared by Commonwealth Mycological Institute. Quarterly (until 1972, twelve issues per volume; from 1973 four issues per volume).
Formerly: An Annotated Bibliography of Medical Mycology.

Arrangement: Arranged by broad subject.

Coverage: Covers selected literature on the diseases of man and animals (domesticated and wild), including fish and some large invertebrates (excluding insects), caused by fungi and actinomycetes; covers their aetiology, symptomatology, epidemiology, and therapy. Fungi-associated allergies of man and animals and poisoning of man and animals by fungi or mould-contaminated feed are also covered. Taxonomic papers are listed in the Institute's half yearly Bibliography of Systematic Mycology and new generic and specific names of fungi are listed in the Index to Fungi.

Scope: International.

Locating Material: Entries are numbered consecutively through each volume.

Abstracts: Abstracts are in English. Titles are in English and in the language of the article except for Japanese, Chinese, Arabic, etc. Cyrillic characters are transliterated according to International Standards. The language of the article and summaries in other languages are noted by an abbreviation. Illustrative material and the number of references are noted. The address of the senior author is given.

Indexes: An author index is in each issue. There is a cumulative subject index for each volume.

Other Material: The Readers' Guide section in the March issue of each year gives detailed information on the scope and use of the Review. It includes "Arrangement of Abstracts" and a list of abbreviations used. The abbreviation list includes those used for Australian states and territories, states of the United States, Canadian provinces, and also those used for languages.

Periodicals Scanned: A list of publications regularly scanned is published from time to time.

Database: Citations are available through CAB ABSTRACTS from 1973 and are updated monthly. Keywords and abstracts are provided. The other abstract journals of the Commonwealth Agricultural Bureaux are also included.

SLEEP BULLETIN, 1968-1980.
Brain Information Service, Brain Research Institute. Ceased.

Arrangement: Arranged by subject.

Coverage: Neurophysiology, physiology, biochemistry, pharmacology, endocrinology, ontogeny, phylogeny, behavior, sleep cycles, dreaming, effect of external stimuli, altered states of consciousness and biofeedback, personality and psychopathology (but not psychoanalytic interpretation of dreams), sleep disorders and deprivation are among the subjects covered. Books, book chapters, reviews, reports, dissertations, and theoretical discussions are included.

Scope: International.

Locating Material: Entries are arranged alphabetically by author within each subject section.

Abstracts: Titles and abstracts are in English. Articles in other languages are indicated by an abbreviation for that language. The senior author's affiliation is given when possible.

Other Material: "Sleep Forum" (articles on various subjects) is found in each issue.

SMALL ANIMAL ABSTRACTS, 1975-
Commonwealth Agricultural Bureaux. Quarterly.

Arrangement: Arranged by broad subject, subdivided into smaller related topics.

Coverage: Bacterial and viral diseases, parasites (ectoparasites, helminths, protozoa) organic diseases, neoplasms, toxicology, immunology, pharmacology, anesthesia, physiology, breeding, genetics, nutrition, biochemistry, haematology, anatomy, radiography, and surgery are among the subjects listed in the Contents.

Scope: International.

Locating Material: Entries are numbered consecutively through each volume. Indexes refer to these numbers.

Abstracts: Titles and abstracts are in English. If the article is in another language, the title is bracketed and also given in that language. An abbreviation which indicates what that language is and summaries in other languages are noted. The senior author's affiliation is provided. The number of references and illustrative material is indicated if this is important.

Indexes: An author index is in each issue which cumulates annually and an annual subject index is provided.

Other Material: A list of abbreviations for journals and for languages is provided with brief explanatory material. A description of Commonwealth Agricultural Bureaux services and products is given.

Database: Only the main abstract journals of the Commonwealth Agricultural Bureaux are listed as being in the CAB ABSTRACTS database. However, entries in the more specialized journals such as Small Animal Abstracts are derived through the same database so they are also found. The earliest entry for any of these is 1973; the update is monthly and keywords and abstracts are provided.

TOXICOLOGY ABSTRACTS, 1978-
Cambridge Scientific Abstracts. Monthly.

Arrangement: Arranged by subject. Broad areas are subdivided into narrower topics.

Coverage: The foreword states that "...the main topics covered are pharmaceuticals, food (including additives and contaminants), agrochemicals, cosmetics, toiletries and household products, industrial chemicals, metals, natural substances (mainly toxins), social poisons and drug abuse, polycyctic hydrocarbons, nitrosamines, and radiation. Toxicology methods and papers concerned   legislation are also included.

Scope: International.

Locating Material: Entries are numbered consecutively through the volume. The letter "T" follows the sequential number and indicates "Toxicology Abstracts," then there is a number for the volume.

Indexes: Author and subject indexes are in each issue. These cumulate annually.

Other Material: A list of abbreviations used in the abstracts and a list of abbreviations for languages appear in each issue. Explanatory material is in the first issue of the volume.

Periodicals Scanned: A list is available on request.

Database: Citations are available through IRL LIFE SCIENCES COLLECTION from 1978. There are monthly updates and abstracts and keywords are provided.

TROPICAL DISEASES BULLETIN, 1912-
Bureaux of Hygiene and Tropical Diseases. Monthly. Published in association with Abstracts on Hygiene.

Arrangement: Arranged by specific disease. The paging is continuous through each issue to the end of the volume.

Coverage: Epidemiology, treatment and management, diagnostic
    methods and manifestations of the diseases covered; venoms
    and anti-venoms, entomology and insecticides, etc. (see
    Table of Contents).

Scope: International.

Locating Material: Entries are numbered consecutively through each
    volume. The Table of Contents and Author Index refer to
    these numbers.

Abstracts: The abstracts vary from quite long and detailed to a
    single sentence. Occasionally no abstract is provided. Titles
    and abstracts are in English. Where feasible the title will
    also be given in the language of the article. One is advised
    if a summary is available in English and of its length.

Indexes: An author and a subject index are provided with each issue.
    Annual author and subject indexes are also published.

Other Material: A list of contributors is found in each issue. Book
    Reviews are usually carried at the end of each issue and
    papers or articles appear from time to time.

Periodicals Scanned: A list appears as a supplement and is published about every two years.

VETERINARY BULLETIN, 1931-
    Commonwealth Agricultural Bureaux, prepared by Commonwealth Bureau of Animal Health. Monthly. Supersedes:
    Tropical Veterinary Bulletin.

Arrangement: Arranged by broad subject.

Coverage: Complete coverage of veterinary medicine including book
    reviews and reports as well as journal articles. Subjects
    listed in the Table of Contents include bacteriology and bacterial diseases, virology and viral diseases, protozoology
    and protozoal diseases; mycoses and mycotoxicosis; anthropod
    and helminth parasites, regional and general pathology; neoplasms and leukosis; nutritional, metabolic and reproductive
    disorders; toxicology, immunology, pharmacology, physiology,
    haematology, anatomy, hygiene, radiations, radioisotopes,
    and surgery.

Scope: International.

Locating Material: Entries are numbered consecutively through each
    volume.

212 / Abstracts and Indexes

Abstracts: Abstracts are in English. Titles are in English and in the language of the article. The language and that of summaries in translation are noted. The senior author's address and the abstractor's initials are provided. Price and ISBN (International Standard Book Number) may be given for monographs. The number of references is indicated.

Indexes: An author index, which cumulates annually, is in each issue. There is an annual subject index.

Other Material: Review articles appear in most issues and a list of abstractors is in each issue. The section "Books Received" is a list of books received recently and being in the list does not preclude that a review of the book will appear. Not included in the Bulletin but available as a separate publication is Veterinary Subject Headings for Use in INDEX VETERINARIUS and the VETERINARY BULLETIN, by R. Mack.

Periodicals Scanned: A list appears in the January issue and includes the name and address of the publisher.

Database: Entries are available through CAB ABSTRACTS since 1972. They are updated monthly and abstracts and keywords are provided. The other abstract journals of the Commonwealth Agricultural Bureaux are also included.

VIROLOGY ABSTRACTS, 1967-
Cambridge Scientific Abstracts. Monthly.

Arrangement: Arranged by broad subjects subdivided into narrower topics.

Coverage: The broad subjects listed in the Contents are: virus taxonomy and classification; physico-chemical properties, structure and morphology; replication cycle; viral genetics including virus reactivation; phage-host interactions including lysogeny and transduction; immunology; antiviral agents; oncology, viral infections of man; diseases associated with slow viruses; viral infections of animals; animal models and experimentally induced viral infections of animals; viral infections of invertebrates; viral infections of fungi and lower plants, and viral infections of higher plants.

Scope: International.

Locating Material: Entries are numbered consecutively through each volume. These numbers contain three elements: the sequential number, a code letter which identifies the abstract journal, and the volume number.

Abstracts: Titles are in English and the language of the article. Up to ten authors are named and the senior author's affiliation is provided. The introductory material contains the statement: "abstracts ... are usually 150 words in length and outline the contents of the paper, the methods used, the results obtained and the conclusions drawn."

Indexes: Author and subject indexes are provided in each issue which cumulate annually.

Other Material: Explanatory material which includes a code for language abbreviations is found in the preliminary pages.

Periodicals Scanned: A list of source journals is available on request.

Database: Entries are available through IRL LIFE SCIENCES COLLECTION from 1978. The updates are monthly and abstracts and keywords are provided.

WILDLIFE DISEASE REVIEW, 1983-
Wildlife Disease Review. Monthly.

Arrangement: Arranged by animal groups (mammals, birds, fish, and reptiles), then taxonomically by species within major groups.

Coverage: Diseases in captive and free-ranging wildlife. An explanatory note states that the publication is "designed to provide current, updated literature to veterinarians, wildlife biologists, animal behaviorists, curators, administrators, researchers, and students...."

Scope: International.

Locating Material: Pages are unnumbered. The entries are numbered sequentially through the issue. The number begins with a digit that indicates the month of issue (1001 is the first entry for January; 2001 is the first entry in February and follows 10044, the last entry for January).

Abstracts: The titles are in English. English abstracts of foreign language articles are provided when possible. The language of an article in a foreign language is noted.

Indexes: Subject, taxonomic, and geographic indexes which are updated monthly are provided. There will be a six-month author index and annual cumulative indexes.

Other Material: Brief explanatory notes are provided. Presently the articles abstracted are from January 1981 to date. There are plans to include literature from 1977-1981 in future issues.

Database: Subject specific searches of literature published after 1960 are available through the publisher.

INDEX

Abstracts in Anthropology 61
Abstracts of Declassified Documents 57
Abstracts of Health Care Management Studies 177
Abstracts of Hospital Management Studies 177
Abstracts of North American Geology 61
Abstracts of World Medicine 178
Abstracts on Hygiene 178, 210
Abstracts on Hygiene and Communicable Diseases 178
Abstracts on Tropical Agriculture 139
ACCESS, List of Periodicals Abstracted by Chemical Abstracts 43
Accumulative Veterinary Index 179
ACM Guide to Computing Literature 25, 29
Acoustics Abstracts 39
Adverse Reactions Titles 188, 190
Aeronautical Engineering 57, 59
Aerospace Abstracts, International 56
Aerospace Medicine and Biology 57, 59, 179
Aerospace Reports, Scientific and Technical 58
Aging, Current Literature on 184
AGRICOLA 145
Agricultural 139
Agricultural and Rural Sociological Abstracts, World 174
Agricultural, Biological and Environmental Sciences 6, 7
Agricultural Economics, American Bibliography of 142
Agricultural Engineering Abstracts 139
Agricultural Index 140, 146
Agricultural Index, Biological and 146
Agricultural Literature of Czechoslovakia 140
Agricultural Sciences 139
Agriculture, Abstracts on Tropical 139
Agriculture, Bibliography of 144
Agrindex 141
AGRIS 142
Air Pollution Abstracts 93
Air Pollution Technical Information Center 93, 94
(Air Transport) Vozdushnyi Transport 18
Air University Library Index to Military Periodicals 1
Air University Periodical Index 1
(Aircraft and Rocket Engines) Aviatsionnye I Raketnye Dvigateli 16
Algology, Mycology, and Protozoology 128

Allegemeine, Angewandte, Regionale und Historische Geologie 76
Allied Health Literature, Cumulative Index to Nursing and 183
Alloys Index 62, 74
Alphabetic Subject Index to Petroleum Abstracts 62
American Bibliography of Agricultural Economics 142
American Botanical Literature, Index to 124
American Doctoral Dissertations 8
Amino Acids, Peptides and Proteins 40, 41, 42, 109
Analysis and Apparatus 39
Anatomy, Anthropology, Embryology and Histology 188
Anesthésie--Réanimation 5
Anesthesiology 188
Anesthesiology Bibliography 195
Animal Abstracts, Small 209
Animal and Human Helminthology 123
Animal Behavior Abstracts 109
Animal Breeding Abstracts 143
Animal Disease Occurrence 179
(Animal Husbandry and Veterinary Science) Zhivotnovodstv I Veterinariya 18
Annalen der Physik (Beiblätter) 49
Annales de Géographie 62
Annual Bibliography of Orthopedic Surgery 195
Antarctic Bibliography 2, 3
Anthropology 61
Anthropology, Abstracts in 61
Apicultural Abstracts 110
Applied Ecology Abstracts 94
Applied Entomology, Review of 131
Applied Mechanics Reviews 79
Applied Science and Technology Index 3
APTIC 93, 94
Aquatic Biology Abstracts 111
Aquatic Sciences and Fisheries Abstracts 111
Archaeological Abstracts, British 67
Archaeology 61, 67
Arid Lands Development Abstracts 144
Art et Archéologie 5
Art-Housing, Furnishings and Equipment 155
Arthritis and Rheumatism 188
Arts and Humanities 6
ASFA 112
Astronomie--Physique Spatiale--Géophysique 4
Astronomischer Jahresbericht 33, 34, 35
Astronomiya (Astronomy) 16
Astronomy 33
Astronomy and Astrophysics Abstracts 33, 34
(Astronomy) Astronomiya 17
Astrophysics Abstracts, Astronomy and 33, 34
Atmospheric Pollution Bulletin 93
Atomes et Molécules, Plasmas 4
Atomindex 10, 55
Automation, New Literature on 30

(Automation, Telemechanics and Computer Technology) Avtomatika, Telemekhanika I Vychislitel' Naya Tekhnika 16
Aviatsio nye I Rake nye Dvigateli (Aircraft and Rocket Engines) 16
Avtomatika, Telemekhanika I Vychislitel' Naya Tekhnika (Automation, Telemechanics and Computer Technology) 16
Avtomobil' nye Dorogi (Motor Roads) 16
Avtomobil' nyi I Gorodskoi Transport (Motor and Municipal Transport) 16

Bacteriology 128
B.A.S.I.C. 115
Bâtiment. Travaux Publics. Transports 5
Behavioural Biology Abstracts 109, 112
Bibliographia Medica 194
Bibliographical Series 25
Bibliographie Astronomique 35
Bibliographie des Sciences de la Terre 4, 62
Bibliographie des Sciences Géologiques 62
Bibliographie Generale de l'Astronomie 35
Bibliographie Géographique Internationale 62
Bibliographie Internationale de Science Administrative 5
Bibliography and Index of Geology 63
Bibliography and Index of Geology Exclusive of North America 63
Bibliography and Index of Micropaleontology 64
Bibliography of Acute Diarrhoeal Diseases 195
Bibliography of Agriculture 144
Bibliography of Astronomy 35
Bibliography of Bioethics 112
A Bibliography of Medical Reviews 195
Bibliography of North American Geology 61, 63, 65, 66
Bibliography of Podiatric Medicine and Surgery 195
Bibliography of Reproduction 113
Bibliography of Scientific and Industrial Reports 9
Bibliography of Seismology 66
Bibliography of Surgery of the Hand 195
Bibliography of Systematic Mycology 132, 208
Bibliography of Technical Reports 9
Bibliography on Cold Regions Science and Technology 3
Bibliography on Medical Education 195
Biochemistry Abstracts 40
Biochimie--Biophysique 4
Bioengineering Abstracts 79
Bioethics, Bibliography of 112
BIOETHICSLINE 113
Biofizika (Biophysics) 16
Biogeography and Climatology 69
Biological Abstracts 106, 114
Biological and Agricultural Index 146
Biological Membrane Abstracts 117
Biological Membranes 40
Biological Sciences 109
Biological Sciences and Living Resources 111

218 / Abstracts and Indexes

Biological Sciences, International Abstracts of 125
Biologie et Physiologie Végetales. Sylviculture 5
Biologiya (Biology) 16
Biology, Aerospace Medicine and 57, 59
(Biology) Biologiya 16
Biophysics, Bio-engineering and Medical Instrumentation 188
(Biophysics) Biofizika 16
BioResearch Index 106, 116
BIOSIS 116, 136, 137
BIOSIS PREVIEWS 116
(Boiler-making) Kotlostroenie 17
Botanica, Excerpta 121
Botanical Literature, Index to American 124
Breeding Abstracts, Animal 143
Breeding Abstracts, Plant 162
British Abstracts 39, 41, 125
British Abstracts of Medical Sciences 125
British Archaeological Abstracts 67
British Geological Literature 67
British Medicine 180
British Society of Rheology Bulletin 49
British Technology Index 80, 81
(Building and Road Machines) Stroitel' nye I Dorozhnye Mashiny 17
Bulletin Analytique 4
Bulletin of Hygiene 178
Bulletin Signalétique 4

CA SEARCH 43
CAB ABSTRACTS 123, 130, 131, 140, 144, 147, 148, 149, 151,
   153, 154, 155, 157, 158, 159, 160, 161, 162, 163, 164, 165,
   166, 167, 168, 169, 171, 172, 174, 180, 197, 208, 210, 212
Cadmium Abstracts 68, 73, 78
Calcified Tissue Abstracts 181
CA-List of Periodicals 43
Cancer 188
Cancer Chemotherapy Abstracts 182
Cancer Therapy Abstracts 182
Carbohydrate Chemistry and Metabolism Abstracts 117
Carbohydrate Metabolism Abstracts 117
Cardiovascular Diseases and Cardiovascular Surgery 188
Cartography, Remote Sensing, Photogrammetry and 69
Cartography, Social Geography and 69
CASSI 42, 43
Catalogue of Medical and Veterinary Zoology, Index 192
CDI 8
Chemical Abstract Service Source Index 42, 43
Chemical Abstracts 41
Chemical Engineering Abstracts, Theoretical 52
(Chemical, Oil-refining and Polymer Machinery) Khimicheskoe,
   Neptepererabatyvayushchee I Polimernoe Mashinostroenie 16
Chemical Reactions, Current 46
Chemisches Centralblatt 43

Chemisches-Pharmaceutisches Centralblatt 43
Chemisches Zentralblatt 43
Chemistry and Index Chemicus, Current Abstracts of 45
Chemistry and Physics 39
(Chemistry) Khimiya 16
Chemo-Reception Abstracts 44
Chemotherapy Abstracts, Cancer 182
Chest Diseases, Thoracic Surgery and Tuberculosis 188
Chimie 4
Chorologica, Taxonomia et 121
Climatology and Hydrology 69
Clinical Biochemistry 188
Clinical Practice 6, 7
COLD REGIONS 3
Cold Regions Science and Technology, Bibliography on 3
Combustibles--Energie 5
Commercial Fisheries Abstracts 117, 127
Committee on Scientific and Technical Information 9, 23, 59
Communicable Diseases, Abstracts of Hygiene and 178
Communication Abstracts 80
Communication Abstracts Journal, Electronics and 83
Community Planning, Regional and 69
COMPENDEX 85
COMPREHENSIVE DISSERTATION INDEX (CDI)
Computer Abstracts 25
Computer and Control Abstracts 19, 20, 26
Computer and Information Systems Abstracts Journal 26
Computer Bibliography 25
Computer News 26
Computer Program Abstracts 27
Computer Rearrangement of Subject Specialties 115
Computer Science 25
Computing Reviews 28
Computing Reviews Reviewer Index 28
Control Abstracts 19
(Coordination of Different Types of Transport Containers)
 Vzaimodeistvie Raznykh Vidov Transport I Konteinernye Perevozki 18
Copley, E. J. 18
Corrosion Abstracts 81
Corrosion and Protection Against Corrosion) Korroziya Zashchita
 ot Korrozii 17
COSATI 9, 23, 59
Cotton and Tropical Fibres Abstracts 146
County Engineers, Journal of the Institution of Municipal and 88
Crane, Eva 111
Cranio-Facial-Cleft Palate Bibliography 195
Crop Physiology Abstracts 147
Crops Abstracts, Field 151
CROSS 115
Cumulated Bibliography and Index to Meteorological and Geoastro-
 physical Abstracts 36
Cumulated Index Medicus 194
Cumulative Index of Hospital Literature 182, 191

220 / Abstracts and Indexes

Cumulative Index to Nursing and Allied Health Literature 183
Cumulative Index to Nursing Literature 183
Current Abstracts of Chemistry and Index Chemicus 45
Current Advances in Genetics 117
Current Advances in Plant Science 118
Current Antarctic Literature 2
Current Bibliography for Aquatic Sciences and Fisheries 111
Current Bibliography for Fisheries Science 111
Current Bibliography of Plastic and Reconstructive Surgery 195
Current Chemical Reactions 46
Current Citations on Strabismus, Amblyopia, and other Diseases of Ocular Mobility 195
Current Contents 6
Current List of Medical Literature 194
Current Literature in Family Planning 184
Current Literature on Aging 184
Current Mathematical Publications 29
Current Physics Index 46
Current Technology Index 81
(Cybernetics) Kibernetika 16

Dairy Science Abstracts 148
Deafness, Speech and Hearing Abstracts (DSH) 187
Deepsea Research and Oceanographic Abstracts 119
Dental Abstracts 185
Dermatologie--Vénéréologie 5
Dermatology and Venereology 188
Descriptive Index of Current Engineering Literature 84
Design Abstracts International 82
Developmental Abstracts 186
Developmental Biology and Teratology 188
Diabète. Obésite. Maladies Métaboliques 5
Disease Occurrence, Animal 179
Disease Review, Wildlife 213
Diseases, Abstracts of Hygiene and Communicable 178
Diseases Bulletin, Tropical 210
Dissertation Abstracts International, Section B: The Sciences and Engineering 7
Dissertations, American Doctoral 8
Dissertations in the Pure and Applied Sciences Accepted by Colleges and Universities in the United States, Master's Theses and Doctoral 14
Distribution Maps of Plant Diseases 132
Doctoral Dissertations, American 8
Doctoral Dissertations in the Pure and Applied Sciences Accepted by Colleges and Universities in the United States, Master's Theses and 14
Documentation Abstracts 12
DOE/RECON 96, 98
Drainage Abstracts, Irrigation and 157
Drug Abuse Bibliography 186
Drug Dependence 188

Drug Literature Index 188, 190
Drugs of Addiction and Non-addiction; Their Use and Abuse: A Comprehensive Bibliography 186
DSH Abstracts 187
Dvigateli Vnutrennogo Sgoraniya (Internal Combustion Engines) 16

Earth Resources 57, 59
Earth Sciences, Archaeology, and Anthropology 61
Ecology Abstracts 94
Economic Geography 69
EDB 96, 98
Ekonomika Promyshlennosti (Industrial Economics) 16
ELCOM 27, 83
ELCOM (Electronics and Computers) 27
Eldoc--Electro--Technique 4
Eldoc--Electronique 4
(Electric Communication) Elektrosvyaz' 16
Electrical and Electronics Abstracts 19, 20, 83
(Electrical and Power Engineering) Elektrotekhnika I Elektroenergetika 16
Electroanalytical Abstracts 47
(Electronic Engineering) Elektronika I EE Primenenie 16
Electronics Abstracts Journal 83
Electronics and Communication Abstract Journal 27, 83
Elektronika I EE Primenenie (Electronic Engineering) 16
Elektrosvyaz' (Electric Communication) 16
Electrotekhnika I Elektroenergetika (Electrical and Power Engineering) 16
EMBASE 190
Endocrinology 188
Energy, A Continuing Bibliography 57, 59
Energy: A Continuing Bibliography with Indexes 95
Energy Abstracts for Policy Analysis 95
Energy Abstracts, Fuel and 100
Energy and Environment 93
Energy Index 96, 97
Energy Information Abstracts 96
ENERGY LINE 97
Energy Research Abstracts 10, 57, 97, 98
Energy Sources, Unconventional 50
Engineering and Technology 79
Engineering Index 84
(Engineering Materials. Construction and Design of Machine Components. Hydraulic Drive) Mashinostroilel'nye Materialy. Konstruktsii I Raschet Detalei Mashin. Gidroprivod 17
Engineering, Technology, and Applied Sciences 6, 7
Entomology Abstracts 120
Entomology, Review of Applied 131
ENVIROLINE 99
Environment 93
Environment Abstracts 98
Environment Information Access 98

Environmental Quality Abstracts 100
Environmental Health and Pollution Control 188
Epilepsy 188
ERDA Energy Research Abstracts 97
ERDA Research Abstracts 57
Excerpta Botanica 121
Excerpta Medica 188

Faba Bean Abstracts 149
Family Economics, Home Management 155
Family Medicine Literature Index 195
Family Planning, Current Literature in 184
Family Relations and Child Development 155
Farm and Garden Index 150
Farmakologiya. Khimioterapevticheskie Sredstva. Toksikologiya (Pharmacology. Chemotherapy. Toxicology) 16
Fertilizer Abstracts 150
Fibres Abstracts, Cotton and Tropical 146
Field Crops Abstracts 151
Fisheries Abstracts, Aquatic Sciences and 111
Fisheries Abstracts, Marine 127
Fisheries Abstracts, World 135
Fisheries Science, Current Bibliography for 111
Fishery Abstracts, Sport 133
Fishery Technological Abstract Card System 127
Fizika (Physics) 16
Food and Nutrition 155
(Food Industry Machinery) Oborudovanie Pishchevoi Promyshlennosti 17
Food Science and Technology Abstracts 152
Forensic Science 188
Forest Products Abstracts 153
Forestry Abstracts 153
Fortschritte der Physik 49
Fotokinotekhnika (Photography and Cinematography) 16
FSTA 153
Fuel and Energy Abstracts 100
Fungi, Index of 132, 208

GAP HYOR, Atomes, Molécules Gaz Neutres et Ionsés 4
Garden Index, Farm and 150
Gas Abstracts 85
Gastroenterology 188
Génie Biomédical--Informatique Biomédical 4
Génie Chimique--Industries Chimique et Parachimique 5
General 1
General Microbiology and Bacteriology 128
General Pathology and Pathological Anatomy 188
General Science Index 8
Genetics Abstracts 122
Genetics, Current Advances in 117

Génétique 5
Geo Abstracts 69
Geoastrophysical Abstracts, Meteorological and 36, 57
Geochemie und Lagerstättenkunde, Petrographie, Technische Mineralogie 77
(Geodesy and Aerial Surveying) Geodeziya I Aeros'emka 16
Geodeziya I Aeros'emka (Geodesy and Aerial Surveying) 16
Geofizika (Geophysics) 16
Geografiya (Geography) 16
Geographical Abstracts 70
Géographique Internationale, Bibliographie 62
(Geography) Geografiya 16
Geologic Literature of North America 65
Geological Abstracts 61
Geological Literature, British 67
Geologie 76
Geologie und Paläontologie, Referate, Neue Jahrbuch für Mineralogie 77
Geologie und Paläontologie, Zentralblatt für 77
Geologisches Zentralblatt 76
Geologiya (Geology) 16
Geology 61, 63, 65, 68, 77
(Geology) Geologiya 16
Geomorphological Abstracts 69
Geophysical Abstracts 70
Geophysics Abstracts 71
Geophysics and Tectonics Abstracts 71
(Geophysics) Geofizika 16
GEOREF 64, 66, 71
GeoRef Serials List and KWOC Index 64
Geo-Science Abstracts 61
Gerontological Abstracts 190
Gerontology and Geriatrics 188
Gornoe Delo (Mining) 16
Gornoe I Neftepromyslovoe Mashinostroenie (Mining and Oil Industry Machines) 16
Government Abstracts, Weekly 10
Government Reports Announcements 9
Government Reports Announcements and Index 9
Government-Wide Index to Federal Research 10
A Guide to Referativnyi Zhurnal 18

Halbmonatliches Literaturverzeichnis 49
Health Aspects of Pesticides Abstract Bulletin 103
Health Care Management Studies, Abstracts of 177
Health Economics and Hospital Management 188
Health Engineering Abstracts, Public 206
Health Literature, Cumulative Index to Nursing and Allied 183
Health Sciences 177
Hearing Abstracts, Deafness, Speech and (DSH) 187
Hearing, Deafness, Speech and 187
Helminthological Abstracts 123

224 / Abstracts and Indexes

Hematology 188
Herbage Abstracts 154
Highway Research Abstracts 85, 90
Highway Research Information Service 87
Highway Safety Literature 86
HIGHWAY SAFETY LITERATURE 87
Histoire des Sciences et des Techniques 5
Histoire et Science de la Littérature 5
Histoire et Sciences des Religions 5
Historische Geologie; Allgemeine, Angewandte, Regionale und 76
Home Economics Research Abstracts 155
Horticultural Abstracts 156
Hospital Abstracts 190
Hospital Literature Index 191, 195
Hospital Management Studies, Abstracts of 177
Houzeau, J.S. 35
HRIS 87, 88
HRIS Abstracts 86, 87, 88
Human and Experimental 159
Human Genetics 188
Hydata 101
Hydrologie--Géologie de l'Ingenieur. Formations Superficielles
    (Bulletin Signalétique. Bibliographie des Sciences de la Terre) 4
Hygiene and Communicable Diseases, Abstracts of 178

IAA 57, 59
IMM Abstracts 72
Immunology Abstracts 124
Immunology, Serology and Transplantation 188
Index Catalogue of Medical and Veterinary Zoology 192
Index Catalogue of the Library of the Surgeon General's Office 194
Index Chemicus 45
Index Medicus 194
Index of Audiovisual Serials in the Health Sciences 195
Index of Fungi 132
Index of Rheumatology 195
Index to American Botanical Literature 125
Index to Dental Literature 195
Index to Fungi 132, 208
Index to Nursing and Allied Health Literature, Cumulative 183
Index to Scientific and Technical Proceedings 10
Index Veterinarius 196
Industrial and Applied Microbiology 128
Industrial Arts Index 3
(Industrial Economics) Ekonomika Promyshlennosti 16
Industrial Microbiology 128
Industrial Reports, Bibliography of Scientific and 9
(Industrial Transport) Promyshlennyi Transport 17
Industries Mécaniques 5
Informatika (Information Sciences) 16
Information and Processing Journal 26
Information Science Abstracts 12

(Information Sciences) Informatika 16
Information Systems Abstracts Journal, Computer and 26
Informatique--Automatique--Recherche Operationnelle--Gestion--
 Economie 4
INIS 56
INIS Automindex: an International Abstracting Service 55
INKA-ASTRO 35
INKA-MATH 32
INSPEC 20
INSPECT INFORMATION SCIENCE 20
Institute of Mining and Metallurgy 72
Institute of Petroleum Abstracts 72
Institution Administration 155
(Internal Combustion Engines) Dvigateli Vnutrennogo Sgoraniya 16
Internal Medicine 188
International Abstracts in Operation Research 12
International Abstracts of Biological Sciences 125
International Aerospace Abstracts 56, 57, 59
International Journal Dealing with the Documentation of All Aspects
 of Fundamental Physico-Chemical and Analytical Electro-Chemistry 47
International Journal of Abstracts 30
International Nuclear Information System 56
International Nursing Index 195, 197
International Petroleum Abstracts 72
International Pharmaceutical Abstracts 198
IPA 199
IRL LIFE SCIENCES COLLECTION 41, 45, 95, 110, 121, 122, 124,
 129, 182, 202, 210, 213
Irrigation and Drainage Abstracts 157
Issledovanie Kosmicheskogo Prostranstva (Space Research) 16

Journal of Current Laser Abstracts 48
Journal of Electroanalytical Chemistry 47
Journal of Helminthology, Supplement 123
Journal of the Institution of Municipal and County Engineers 88
Journal of the Mineralogical Society 75
Journal of the Textile Institute 175

Keyword Index of Wildlife Research 126
Khimicheskoe, Neftepererabatyvayushchee I Polimernoe Mashinostroenie
 (Chemical, Oil-refining and Polymer Machinery) 16
Khimiya (Chemistry) 16
Kibernetika (Cybernetics) 16
Kommunal'noe, Bytovoe I Torgovoe Oborudovanie (Municipal, House-
 hold and Trading Equipment) 17
Korroziya I Zashchita ot Korrozii (Corrosion and Protection Against
 Corrosion) 17
Kotlostroenie (Boiler-making) 17
Kristallographie und Mineralogie 77

226 / Abstracts and Indexes

Lagerstättenkunde; Petrographie, Technische Mineralogie, Geochemie und 77
Lelande, J.J. de 35
Lancaster, A. 35
Land Use Planning Abstracts 102
Landforms and the Quarternary 69
Lands Development Abstracts, Arid 144
Laser Abstracts, Journal of Current 48
Laser and Electro-Optic Reviews 50
Laser/Maser International 48
Lead Abstracts 68, 73, 78
Legkaya Promyshlennost' (Textile Industry) 17
Leprosy and Related Subjects 188
Leukemia Abstracts 199
Life Sciences 6
List of References from Current Literature 148
List of References on Nuclear Energy 55
List of Translations Notified to the International Translations Centre 23
Literature in Family Planning, Current 184
Literature on Aging, Current 184
Livestock Feeds and Feeding 159

MAB 95
Magnetohydrodynamics and Plasmas 50
Maize Quality and Protein Abstracts 158
Maladies de l'Appareil Digestif--Chirurgie Abdominale 5
Maladies de l'Appareil Respiratoire du Coeur et des Vaisseaux-- Chirurgie Thoracique et Vasculaire 5
Maladies des Os et des Articulations--Chirurgie Orthopédique-- Traumatologie 5
Maladies des Reins et des Voies Urinaires--Chirurgie de l'Appareil Urinaire 5
Maladies du Sang 5
Maladies du Système Nerveaux--Myopathies--Neurochirurgie 5
Man and Biosphere 95
Marine Fisheries Abstracts 127
Mashinostroitel'nye Materialy. Konstruktsii I Raschet Detalei Mashin. Gidroprivod (Engineering Materials. Construction and Design of Machine Components. Hydraulic Drive) 17
Mass Transportation Abstracts, Urban 91
Master's Theses Accepted by U.S. Colleges and Universities in the Fields of Chemical Engineering, Chemistry, Mechanical Engineering, Metallurgical Engineering, and Physics 14
Master's Theses and Doctoral Dissertations in the Pure and Applied Sciences Accepted by Colleges and Universities in the United States 14
Master's Theses in the Pure and Applied Sciences Accepted by Colleges and Universities in the United States and Canada 14
Matematika (Mathematics) 17
Mathematic und ihre Grenzgebiete, Zentralblatt für 31
Mathematical Reviews 29, 30
(Mathematics) Matematika 17

Mathematics, Statistics, and Computer Science 25
MATHFILE 30
(Mechanical Engineering) Tekhnologiya Mashinostroeniya 17
(Mechanics) Mekanika 17
Mechanics Reviews, Applied 79
Medica, Excerpta 188
Medical and Veterinary 131
Medical and Veterinary Mycology, Review of 207
Medical and Veterinary Zoology, Index Catalogue of 192
(Medical Geography) Meditsinskaya Geografiya 17
Medical Subject Headings Annotated Alphabetic List 194
Medicine, Abstracts of World 178
Medicine and Biology, Aerospace 57, 59
Medicine, British 180
Medicine, Physical Fitness/Sports 204
Medicus, Index 194
Meditsinskaya Geografiya (Medical Geography) 17
Meditsinskii Referativnyi Zhurnal 200
MEDLARS 196, 198, 204
MEDLINE 196, 198, 204
Mekanika (Mechanics) 17
Mental Health Book Review Index 200
Mental Retardation Abstracts 186, 201
Metabolism Abstracts, Carbohydrates Chemistry and 117
METADEX 74
Metal Literature, Review of 74
Metallurgical Abstracts 74
Metallurgiya (Metallurgy) 17
Metallurgy, Institute of Mining and 72
(Metallurgy) Metallurgiya 17
Metals Abstracts 74
Métaux-Métallurgie 5
Meteorological Abstracts and Bibliography 36
Meteorological and Geoastrophysical Abstracts 36
Metrologiya I Zmeritel'naya Tekhnika (Metrology and Measuring Instruments) 17
(Metrology and Measuring Instruments) Metrologiya I Zmeritel'naya Tekhnika 17
MGA 37
Microbiologie--Virologie--Immunologie 5
Microbiology 188
Microbiology Abstracts 128
Microfilm Abstracts 7
Micropaleontology, Bibliography and Index of 64
Microscopie Electronique--Diffraction Electronique 5
Military Periodicals, Air University Library Index to 1
Millets Abstracts, Sorghum and 168
Mineralogical Abstracts 75
Mineralogical Magazine, Supplement to 75
Mineralogical Society, Journal of the 75
Minéralogie--Géochimie--Géologie Extraterrestre (Bulletin Signalétique. Bibliographie des Sciences de la Terre) 4
Mineralogie, Geologie und Paläontologie, Referate; Neues Jahrubch für 77

228 / Abstracts and Indexes

Mineralogie, Zentralblatt für   77
Mining and Metallurgy, Institute of   72
(Mining and Oil Industry Machines) Gornoe I Neftepromyslovoe Mashinostroenie   16
(Mining) Gornoe Delo   16
Monthly Bibliography of Medical Reviews   201
(Motor and Municipal Transport) Avtomobil'nyi I Gorodskoi Transport   16
(Motor Roads) Avtomobil'nye Dorogi   16
MRIS Abstracts   88
Multiple Sclerosis Indicative Abstracts   201
Municipal and County Engineers, Journal of the Insititution of   88
(Municipal, Household and Trading Equipment) Kommunal'noe, Bytovoe I Torgovoe Oborudovanie   17
Mycology, and Protozoology; Algology   128
Mycology, Bibliography of Systematic   208
Mycology, Review of Applied   131
Mycology, Review of Medical and Veterinary   207

NAPCA Abstract Bulletin   93
NASA   57, 59
NASA Patent Abstracts   57, 59
NASA patents   57
Nasosostroyeniye I Kompressorostroenie. Kholodil'noe Mashinostroenie (Pumps and Compressors. Refrigeration)   17
National Library of Medicine Current Catalog   196, 202
Neues Jahrbuch für Mineralogie, Geologie und Paläontologie, Referate   77
Neurology and Neurosurgery   188
Neurosciences Abstracts   202
Neurosurgical Biblio Index   195
New Literature on Automation   30
NTIS   10, 87, 91
NTISEARCH   10
Nuclear Medicine   188
(Nuclear Reactors) Yadernye Reaktory   18
Nuclear Science Abstracts   10, 55, 57
Nuclear Science and Space Science   55
Nucleic Acids   40
Nucleic Acids Abstracts   49, 129
Nuisances   5
Nursing and Allied Health Literature, Cumulative Index to   183
Nursing Index, International   197
Nutrition Abstracts and Reviews   159
Nutrition Planning   160

Oborudovanie Pishchevoi Promyshlennosti (Food Industry Machinery)   17
Obshchie Voprosy Patologii (General Problems of Pathology)   17
Obstetrics and Gynecology   188
Occupational Health and Industrial Medicine   188
Ocean Technology, Policy and Nonliving Resources   111
Oceanic Abstracts   129

OCEANIC ABSTRACTS 130
Oceanic Citation Journal 129
Oceanic Index 129
Offshore Abstracts 88
Oil Seeds Abstracts, Tropical 172
Okhrana Prirody I Vosproizvodstvo Prirodnykh Resursov 17
(Oncology) Onkologiya 17
Onkologiya (Oncology) 17
Operation Research, International Abstracts in 12
Ophtalmologie 5
Ophthalmology 188
Oral Research Abstracts 202
Organizatsiya I Bezopasnost' Dorozhnogo Dvizheniya 17
Organizatsiya Upravleniya (Industrial Management and Organization) 17
Ornamental Horticulture 160
Orthopedic Surgery 188
Oto-Rhino-Laryngologie--Stomatologie--Pathologie Cervicofaciale 5
Otorhinolaryngology 188

Paläontologie 76
Paläontologie, Referate; Neues Jahrbuch für Mineralogie, Geologie und 77
Paläontologisches Zentralblatt 77
Paläontologie, Zentralblatt für Geologie und 77
Paléontologie (Bulletin Signalétique. Bibliographie des Sciences de la Terre) 4
Pandex Current Index to Scientific and Technical Literature 14
Parasite--Subject Catalogues 192
Patent Abstracts, NASA 57, 59
patents, NASA 57
Pediatrics and Pediatric Surgery 188
Peptides and Proteins; Amino Acids 40, 41, 42, 109
Pesticides Abstracts 103
Pesticides Documentation Bulletin 104
Petrographie, Technische Mineralogie, Geochemie und Lagerstättenkunde 77
Petroleum Abstracts 75
Petroleum Abstracts, International 72
Pharmaceutical Abstracts, International 198
Pharmaceutisches Centralblatt 43
Pharmacology and Toxicology 188
(Pharmacology. Chemotherapy. Toxicology) Farmakologiya. Khimioterapevticheskie Sredstva. Toksikologiya 16
Philosophie 5
Photogrammetry and Cartography, Remote Sensing 69
(Photography and Cinematography) Fotokinotekhnika 16
Physical, Chemical and Earth Sciences 6
Physical Education Index 203
Physical Fitness/Sports Medicine 195, 204
Physics 39
Physics Abstracts 19, 49
Physics and Electrical Engineering (Science Abstracts) 19

(Physics) Fizika 16
Physics Index, Current 46
Physikalische Berichte 49
Physiology 188
Physique, Chimie et Technologie Nucleaires 4
Physique de l'Etat Condensé 4
Physique Mathématique, Optique, Acoustique, Méchanique, Chaleur 4
Pig News and Information 161
(Pipelines) Truboprovodnyi Transport 17
Plant Breeding Abstracts 162
(Plant Breeding) Rastenievodstvo 17
Plant Diseases, Distribution Maps of 132
Plant Growth Regulator Abstracts 163
Plant Nematology 123
Plant Pathology, Review of 132
Plant Science, Current Advances in 118
Plastic Surgery 188
Pochvovedenie I Agrokhimiya (Soil Science and Agricultural Chemistry) 17
Policy Analysis, Energy Abstracts for 95
Pollination of Seed Crops 111
Pollution Abstracts 104
POLLUTION ABSTRACTS 105
Polymères. Peintures. Bois. Cuirs 5
Potato Abstracts 164
Poultry Abstracts 165
Pozharnaya Okhrana (Protection Against Fire) 17
Préhistoire et Protohistoire 5
(Problems of Technical Progress and Management in Engineering)
  Voprosy Teknicheskogo Progressa I Organizatsiya Proizvodstva
  V Mashinostroenii 17
Proceedings in Print 15
Produits Alimentaires 5
Promyshlennyi Transport (Industrial Transport) 17
(Protection Against Fire) Pozharnaya Okhrana 17
Protein Abstracts, Maize Quality and 158
Proteins; Amino Acids, Peptides and 40, 41, 42, 109
Protozoaires et Invertébrés. Zoologie Générale et Appliquée 5
Protozoological Abstracts 130
Protozoology; Algology, Mycology and 128
PSYCHABS 205
Psychiatry 188
Psychological Abstracts 204
Psychologie--Psychopathologie--Psychiatrie 5
Psychopharmacology Abstracts 206
Psychopharmacology Bibliography 195
Public Health Engineering Abstracts 206
Public Health, Social Medicine and Hygiene 188
(Pumps and Compressors. Refrigeration) Nasosostroyeniye I Kompressorostroenie. Kholodil'noe Mashinostroienie 17

Quantum Electronics 50
Quarterly Bibliography of Major Tropical Diseases 195

(Radiation Biology) Radiasionaya Biologiya 17
Radiatsionaya Biologiya (Radiation Biology) 17
(Radio Engineering) Radiotekhnika 17
Radiology 188
Radiotekhnika (Radio Engineering) 17
(Rail Transport) Zheleznodorozhnyi Transport 18
Raketostroenie (Rocket Engineering) 17
Rastenievodstvo (Plant Breeding) 17
Recurring Bibliography of Hypertension 195
Recurring Bibliography on Education in the Allied Health Professions 195
Referativnyi Zhurnal 16
Rehabilitation and Physical Medicine 188
Rehabilitation Literature 207
Remote Sensing, Photogrammetry and Cartography 69
Reproduction, Bibliography of 113
Reproduction. Embryologie. Endocrinologie 5
Research Abstracts 90
Research Reports, U. S. Government 9
Review of Applied Entomology 131
Review of Applied Entomology; Series A, Agricultural 165
Review of Applied Entomology; Series B, Medical and Veterinary 207
Review of Applied Mycology 132
Review of Medical and Veterinary Mycology 207
Review of Metal Literature 74
Review of Plant Pathology 132
Revue Bibliographique Cancer (editée par l'Institut Gustave Roussy) 5
Rheology Abstracts, A Survey of World Literature 49
Rice Abstracts 166
Richter, M. M. 44
Regional and Community Planning 69
Regionale und Historische Geologie; Allgemeine, Angewandte 76
Road Abstracts 88
Roches Cristallines (Bulletin Signalétique. Bibliographie des Sciences de la Terre) 4
Roches Sedimentaires--Géology Marine (Bulletin Signalétique. Bibliographie des Sciences de la Terre) 4
(Rocket Engineering) Raketostroenie 17
RRIS Bulletin 88
Rural Sociological Abstracts, World Agricultural Economics and 174

Safety Science Abstracts Journal 89
SAFETY SCIENCE ABSTRACTS JOURNAL 90
SCI 21
SCISEARCH 7, 21
Science Abstracts 19
Science Abstracts, Physics and Electrical Engineering 19
Science Citation Index 7, 20
Science de l'Information--Documentation 4
Science du Langage 5
Science Research Abstracts Journal 50
Sciences Agronomiques. Produits Végétales 5
The Sciences and Engineering, Dissertation Abstracts, Section B: 7

232 / Abstracts and Indexes

Sciences de l'Education 5
Sciences Pharmacologiques--Toxicologie 5
Scientific and Industrial Reports, Bibliography of 9
Scientific and Technical Aerospace Reports 10, 57, 58, 59
Scientific and Technical Literature, Pandex Current Index to 14
Scientific and Technical Proceedings, Index to 10
Seed Abstracts 166
Seeds Abstracts, Tropical Oil 172
Seismology, Bibliography of 66
Selected Water Resources Abstracts 105
Semiconductor Electronics 51
Serials Sources for BIOSIS Data Base 116
SHE 84
SIS/ISTP&B 12
Sleep Bulletin 208
Small Animal Abstracts 209
Social and Behavioral Sciences 6
Social and Historical Geography 69
Social Geography and Cartography 69
Social Science Citation Index 21
Sociologica 121
Sociology Abstracts, World Agricultural Economics and Rural 174
Sociologie--Ethnologie 5
(Soil Science and Agricultural Chemistry) Pochvovedenie I Agrokhimiya 17
Soils and Fertilizer 167
Solid State Abstracts 51
Solid State Abstracts Journal 51
Solid State Abstracts on Cards 52
Sorghum and Millets Abstracts 168
Soudage, Brasage et Techniques Connexes 5
Soyabean Abstracts 169
(Space Research) Issledovanie Kosmicheskogo Prostranstva 16
Space Science 55
Spectrochemical Abstracts 52
Speech and Hearing Abstracts, Deafness (DSH) 187
Speech and Hearing, Deafness 187
SPIN 47
Sport Fishery Abstracts 133
/Sports Medicine; Physical Fitness 204
STAR 57, 59
Statistical Theory and Method Abstracts 30
Statistics 25
Stratigraphie--Géologie Régionale et Géologie Générale (Bulletin Signalétique. Bibliographie des Sciences de la Terre 4
Stroitel'nye I Dorozhnye Mashiny (Building and Road Machines) 17
Structure de l'Etat Condensé. Cristallographie 4
Subject Index to Periodicals 81
Summary of Current Literature 107, 175
Superconductivity 50
Supplement to Mineralogical Magazine 75
Surgery 188
Svarka (Welding) 17

Taxonomia et Chorologica 121
Technical Book Review Index 21
Technical Literature, Pandex Current Index to Scientific and 14
Technical Proceedings, Index to Scientific and 10
Technical Publications Announcements 57
Technical Reports, Bibliography of 9
Technical Translations 9
Technology 79
(Technology and Machinery of Paper-making and Printing) Tekhnologiya I Oborudovanie Tsellyulozno-Bumazhnogo I Poligraficheskogo Proizvodstva 17
Technology Index, Current 81
Tectonics Abstracts, Geophysics and 71
Techtonique (Bulletin Signalétique. Bibliographie des Sciences de la Terre) 4
Tekhnologiya I Oborudovanie Tsellyulozno-Bumazhnogo I Poligraficheskogo Proizvodstva (Technology and Machinery of Paper-making and Printing) 17
Tekhnologiya Mashinostroeniya (Mechanical Engineering) 17
Teploenergetika (Thermal Power Engineering) 17
Textile Abstracts 175
Textile Abstracts, World 175
(Textile Industry) Legkaya Promyshlennost' 17
Textile Institute, Journal of the 175
Textile Technology Digest 170
Textile and Clothing 155
Theoretical Chemical Engineering Abstracts 52
Theoretical Physics 50
(Thermal Power Engineering) Teploenergetika 17
Tissue Culture Abstracts 133
Tobacco Abstracts 170
Torrey Botanical Club Bulletin 124
Toxicology Abstracts 210
TOXLINE 199
(Tractors and Farm Machinery and Equipment) Traktory I Sel'skokhozyaistvennye Mashiny I Orudiya 17
Traktory I Sel'skokhozyaistvennye Mashiny I Orudiya (Tractors and Farm Machinery Equipment) 17
Translations Register Index 22
Transportation Research Abstracts 85, 90
Transportation Research Information Service 86
TRIS 86, 88, 91
Triticale Abstracts 171
TROPAG 139
Tropical Abstracts 139, 172
Tropical Agriculture, Abstracts on 139
Tropical Diseases Bulletin 210
Tropical Fibres Abstracts, Cotton and 146
Tropical Oil Seeds Abstracts 172
Tropical Veterinary Bulletin 211
Truboprovodnyi Transport (Pipelines) 17
TULSA 76
(Turbine Engineering) Turbostroenie 17

234 / Abstracts and Indexes

Turbostroenie (Turbine Engineering) 17

Unconventional Energy Sources 50
U. S. Geological Survey Bulletin 65, 66
U. S. Government Research and Development Reports 9
U. S. Government Research Reports 9
Urban Mass Transportation Abstracts 91
Urology and Nephrology 188

Veterinarius, Index 196
The Veterinary Bulletin 196, 212
Veterinary Index, Accumulative 179
Veterinary Mycology, Review of Medical and 207
Veterinary Zoology, Index Catalogue of Medical and 192
Virology 188
Virology Abstracts 212
Vodnyi Transport (Water Transport) 17
Voprosy Teknicheskogo Progressa I Organizatsiya Proizvodstva V Mashinostroenii (Problems of Technical Progress and Management in Engineering) 17
Vozdushnyi Transport (Air Transport) 18
Vzaimodeistvie Raznykh Vidov Transport I Konteinernye Perevozki (Coordination of Different Types of Transport. Containers) 18

Water Pollution Abstracts 107
Water Pollution Research, Summary of Current Literature 107
Water Resources Abstracts, Selected 105
(Water Transport) Vodnyi Transport 17
Weed Abstracts 173
Weekly Government Abstracts 10
(Welding) Svarka 17
Wildlife Abstracts 134
Wildlife Disease Review 213
Wildlife Research, Keyword Index of 126
Wildlife Review 134
World Agricultural and Rural Sociological Abstracts 174
World Agricultural Economics Abstracts 174
World Agricultural Economics and Rural Sociology 142
World Fisheries Abstracts 135
World Index of Scientific Translations 23
World Medicine, Abstracts of 178
World Textile Abstracts 175
WORLD TEXTILES 176
World Transindex 23
WRA 106
WTI 23

Yadernye Reaktory (Nuclear Reactors) 18

Zeitschrift für Angewandte Chemie   43
Zentralblatt für Geologie und Paläontologie   77
Zentralblatt für Mathematic und ihre Grenzgebiete   31
Zentralblatt für Mineralogie   77
Zheleznodorozhnyi Transport (Rail Transport)   18
Zhivotnovodstvo I Veterinariya (Animal Husbandry and Veterinary Science)   18
Zinc Abstracts   68, 73, 78
ZLC ABSTRACTS   69, 73, 78
Zoological Record   136
ZOOLOGICAL RECORD   137
Zoologie des Vertebres, Ecologie Animale. Physiologie Appliqué Humaine   5
Zoology, Index Catalogue of Medical and Veterinary   192

**LIBRARY USE ONLY
DOES NOT CIRCULATE**